The Philosophy of (Artificial) Intelligence

Ermanno Bencivenga

The Philosophy of (Artificial) Intelligence

Building a New Perspective from the Ground Up

Springer

Ermanno Bencivenga
Distinguished Professor of Philosophy
and the Humanities, Emeritus
University of California
Irvine, CA, USA

ISBN 978-3-032-09346-2 ISBN 978-3-032-09347-9 (eBook)
https://doi.org/10.1007/978-3-032-09347-9

This Springer imprint is published by the registered company Springer Nature Switzerland AG
The registered company address is: Gewerbestrasse 11, 6330 Cham, Switzerland

Prologue

Artificial intelligence is all over the map. It is, increasingly, running our companies, scribbling our texts, fighting our wars. Many are thrilled, including those who expect to make fabulous profits from it, and many more are worried, as they fear they might be losing their jobs, maybe their souls. Since I have been writing about this topic for a long time, in works that have so far attracted little attention, I believe I should give it one last try, now that the topic has become, indeed, topical, and do so in a patient and articulate way, bringing out all the multiple elements of a picture which is exceedingly complex.

One main reason why my work has attracted little attention is connected to the parenthesis in my title. People writing on AI have always been anxious to show their technical competence in the field, and have behaved like those who flaunt the jargon of musical criticism without being able to read music. I have always thought, on the contrary, that we need to go to the root of things: that an understanding of artificial intelligence, of its potential and its threat, is a simple corollary of an understanding of intelligence, period, and of such related concepts as language, thought, consciousness, and will. Therefore here, too, I only get late to the intelligence of machines, and then briefly. I make remarks about it along the way, to keep the reader aware of where all of it is going; but, once we get to the destination, nothing much more is left than appreciating the lessons we have been learning, and summarizing their implications.

The implications are, I am afraid, rather bleak for those rooting for *homo sapiens*, who has been forfeiting his *sapientia* for quite some time, just as he was enriching the *sapientia* of some of his own creations (not optimally yet: AI insists on correcting my "*sapientia*" into "*sapiential*"). I cannot say I am happy about this prospect: I am a member of that forsaken species and, despite all the evil it has unleashed, I feel sorry for it. But, despite my personal feelings, and as I say at the very end, I am not sure that its impending demise is not for the better.

Since I first wrote on this topic, and even before, I also worked on a variety of other issues, and much of that turns out to be relevant to my current concerns. Don't be surprised, then, if I make numerous references to, and even quotes from, such other work of mine—whenever I believe it can help the project at hand.

Brooklyn, New York, USA
September 2025

Contents

Chapter 1
Play

Abstract A behavior is a primitive here, so it cannot be defined, though examples can be given of it. A behavior has a subject (what behaves), which is an internal parameter of it (ontologically dependent on it), and has (or can have) other parameters as well: a space where, and a time when, it occurs; something it applies to. The same behavior can recur; it can expand and contract; and it can change. To the extent that it does not change, it is steady; to the extent that it changes, it plays. And it can learn from its play: so evolve as to face difficulties that have arisen, but also to get into more difficulties (helplessness, and self-destructiveness, can be learned). Behaviors and their subjects can interact, causing change and play in some or all of them, which can be productive for one while it is unproductive, or even counterproductive, for another. A subject that was absolutely steady would be an object. But there can be no objects: every subject eventually changes, and plays; it can (indeed, will) even return from the dead. What there is is always a compromise between play and objectivity; and the amount of objectivity that is present allows for the practice of scientific disciplines. Whose predictions hold until they no longer do: Kant taught us that we live in a world of phenomena, which may be well-behaved within a limited context (and for a while) but are bound to explode if we try to take a universal point of view on them. A more contemporary jargon would reformulate Kant's antinomies in terms of chaos theory. In a chaotic world, play is the engine of survival, and there is a constant struggle for who gets to play, and to survive.

Keywords Behavior · Subject · Play · Learning · Phenomena · Survival

A *behavior* is a primitive here, so it cannot be defined.[1] I can give examples of it. The rolling of a cloud across the sky is a behavior; so are a cat purring, a tree growing leaves, and a sprinter crossing the finish line. As the examples make clear, in behaviors something behaves (the cloud, the cat, the tree, the sprinter)[2]; but that is

---

[1] The notion of a primitive will come up for discussion in Chapter 8.

[2] It would seem that this is not true of all behaviors: for example, that nothing rains or snows when "it" rains or snows. But then I would want to say that "the weather" rains or snows. (In Book XIV of the *Odyssey*, lines 457–458, we read an anthropomorphic version of it: "Zeus rained all night

E. Bencivenga, *The Philosophy of (Artificial) Intelligence*,
https://doi.org/10.1007/978-3-032-09347-9_1

not to be understood as an autonomous entity to which the behavior is attached, as in the Aristotelian paradigm where qualities and quantities are attached to substances, and ontologically dependent on them—if the substance *apple* ceases to exist, so does its quality of *red*. What behaves is rather an internal parameter of the behavior (the ontological dependence is reversed from Aristotle), which distinguishes it from similar behaviors: the rolling of a cloud is distinct from that of a wave, and the purring of a cat is distinct from that of a lynx, in the same way in which the picture of a man is distinct from the picture of a woman or the thought of a unicorn is distinct from that of a mermaid, whether or not there are any men or women to be depicted, any unicorns or mermaids to be thought of.

There may be other (internal) parameters to a behavior. A space where it occurs: the sky for the cloud rolling, a house kitchen for the cat purring, a forest for the tree growing leaves, an arena for the sprinter crossing the finish line. A time when it takes place: a cloud can roll in the sky for an hour; a cat can purr for a few minutes; a tree can grow leaves through a spring; a sprinter can cross the finish line in a split second. And there may be something the behavior applies to: the tree's applies to the leaves, the sprinter's applies to the finish line; the cloud's and the cat's don't seem to apply to anything. Just as what is behaving, these other parameters have no autonomy from the behaviors they belong to: the sky where the cloud rolls is no more detachable from the cloud's rolling behavior than the Land of Oz from the story (or the musical) where it is described, nor are (from the relevant behaviors) the kitchen where the cat purrs, the forest where the tree grows leaves, or the arena where the sprinter crosses the finish line; the times pertinent to all these behaviors can no more be inserted in a unique, objective, chronological time than those pertinent to Aeneas' adventures as narrated by Virgil; you are not going to find the finish line crossed by the sprinter in a catalogue of implements foreign to that particular crossing.

Nothing rules out, in principle, behaviors that have no spatial o temporal parameters. A natural number being added to another is a behavior, and so is the Platonic idea of the Good being "the cause of the knowledge and truth."[3] But here I choose to only consider behaviors that occur in spacetime; hence "behavior" will be equivalent, from now on, to "spatiotemporal behavior."

A behavior can recur (reoccur, be repeated). There can be multiple occurrences of a cloud rolling, of a cat purring, of a tree growing leaves, of a sprinter crossing the finish line. Some of them may just resemble one another: be multiple, distinct examples of *clouds rolling*, *cats purring*, *trees growing leaves*, *sprinters crossing the finish line*. And some of them may be occurrences of the very same behavior: the same rolling of a cloud in the sky, the same purring of a cat, the same growing of leaves by a tree, the same crossing of a finish line by a sprinter. Because behaviors are primitive here, so are their conditions of identity: it is just a brute fact that sometimes we face identical behaviors and at other times distinct ones.

---

long." See it at Lattimore [1967], 222.) Also, as will become apparent later, in some behaviors more than one thing behaves.

[3] Plato (1991, 189 (508d)).

The same behavior can expand in scope—spatial, temporal, or otherwise—from one occurrence to another. A cloud's rolling in the sky can expand from an hour to a whole afternoon: the cloud is the same because it is the same parameter of (numerically) the same behavior; the time when it takes place expands. A cat's purring in the kitchen can expand to its purring all over the house; a tree growing leaves through a spring can expand to it also growing branches; a sprinter crossing the finish line can expand from an arena to the world's stage. Or, moving in the opposite direction, a behavior can recur contracted in scope (again, relative to any of its parameters).

When a behavior's scope expands, the behavior can change from one occurrence to another.[4] The cloud that rolled intact in the sky can, in an expanded scope, no longer roll intact but break in two; the cat that was quietly purring can, in an expanded scope, no longer quietly purr but get scared and start running away; the tree that was growing leaves can, in an expanded scope, get stunted; the sprinter that was crossing the finish line can, in an expanded scope, collapse before (the split second before) he reaches it. When there is no change between two occurrences of a behavior we say that they are (and the behavior is) *steady*; when there is change between them we say that there is *play* going on (and the behavior plays).

Two behaviors can interact. That is: one of them, or both, can expand in scope so as to include the other, and within the expanded scope one of them, or both, can behave so as to impact the other. A cloud rolling in the sky can collide with a second cloud also rolling there, issuing in them merging, and maybe in lightning and thunder; in the resulting behavior (expanded from the previous one, or ones), what behaves is both clouds. A purring cat can be picked up and held by its owner, and happily keep purring in her lap; in the resulting behavior, what behaves is both the cat and the owner. A tree growing leaves can be chosen as the seat for a bird's nest; in the resulting behavior, what behaves is both the tree and the bird (or birds). A sprinter crossing the finish line can be blinded by the sunlight and stumble just slightly; in the resulting behavior, what behaves is both the sprinter and the sunlight. As, again, the examples make clear, change in a behavior may be (though it is not necessarily) the outcome of interaction: a cloud can break because of a rising temperature in the sky; a cat can interrupt its purr and run away because a dog scares it; a tree can become stunted because it becomes infested with parasites; a sprinter can collapse before the finish line because a competitor trips him up. But change can also occur without interaction: a cloud can dissipate because it just runs out of steam; a cat can stop purring because it falls asleep.

What behaves in a behavior is a *subject*. The cloud is a subject of its rolling in the sky, as well as of its merging with another cloud, and of bringing about lightning

[4] I am happy if this notion of change is understood intuitively. If we were to make it precise, we might be tempted to say that a behavior changes when it breaks the rule it was following; but then we would face the Wittgensteinian paradox: "This was our paradox: no course of action could be determined by a rule, because every course of action can be made out to accord with the rule" (Wittgenstein 1953, 81). So I prefer to say that a behavior changes when it takes a course different from what one was probabilistically led to expect. (In case you wonder, I favor a subjective interpretation of probability, but it doesn't really make any difference.)

and thunder; so is the other cloud, to the extent that it is involved. The cat is a subject of its purring, as well as of its being picked up and held by its owner; so, to the extent that she is involved, is the owner. The tree, and the bird or birds, are subjects of the various occurrences of the growing of leaves; the sprinter, and the sunlight, and the competitor tripping him up, of the various occurrences of the crossing (or not crossing) the finish line. As a behavior expands, contracts, or changes, so do its subjects: the cloud expands as it rolls through a whole afternoon (or contracts as it dissipates); the cat changes as it runs away scared (or falls asleep); the tree contracts as leaves no longer grow on it; the sprinter changes as he gets tripped up; the *two* clouds contract as, after merging, they dissolve into rain. But, when there are more than one subject to a behavior, there are also more than one behavior present, each seen from the point of view of one of the subjects. The points of view can be entirely symmetrical, and peaceful, so that it makes no difference which of them we privilege: the two clouds merging can be an indistinguishable phenomenon whether it be the first cloud merging with the second one or vice versa. But they can also be, to various extents, diverging and antagonistic. To give a less innocuous example than any of the above, a teacher enforcing discipline on a pupil may be quite a different behavior if seen as the teacher's or the pupil's behavior: from the former point of view, it can be a case of successful education; from the latter, of equally successful repression.[5]

A subject whose behavior, no matter how far it expanded, always had a further expansion that showed it to be steady would be an *object*. A billiard ball hit several times with a stick may on one occasion move along a surprisingly new trajectory; but when the scope of this behavior is expanded to take careful note of the exact point where the ball was hit, and hence of the spin that was put on it, and many other cases of hitting the ball, with or without spin, are included in this expanded behavior, it will appear that there has been no change whatsoever—the ball has always moved along trajectories that can be calculated with mathematical precision. And calculated, at least in principle, in advance, to prove that nothing new is going on: the behavior is unchangeably tied to an inflexible repetition of before and after. The ball would be an object, in other words, from whose behavior no change can be expected: it would be completely steady.

Despite the promises held by this last example, the use of the phrase "would be" a few times suggests that, in speaking of objects, we are speaking of nonentities. It is not only that there aren't any: there *can be* none; we are spelling out an impossible concept, as if we tried the same with a round square, or a barber who shaves all

---

[5] A brief remark is in order, for those who would want to compare what I say here to some of my other work—for example, my (1997, 2013, 2023). There my starting point is patterns, and subjects are the locus for the encounter of patterns: the locus of difference. Here I move at a more limited, more local, shallower metaphysical level, more in contact with the intended object of my inquiry: (artificial) intelligence. So I start with (spatiotemporal) behaviors, which result from the encounter of patterns (in spacetime), hence contain difference already—internal difference that can increase when they interact with other behaviors. What acts that difference, what performs it (in spacetime), is still a subject; but, since behaviors are what a subject would act, there is perfect correspondence between behaviors and subjects. And I choose to exploit this correspondence by stressing the primacy of behaviors, making subjects the subjects *of behaviors*, in order to put the notion of behavior front and center right away (it will play a crucial role in what follows).

and only those who do not shave themselves. This is the first memorable result of Kant's critical inquiry: in order to grant the steadiness of even a single object, a lowly billiard ball, we *would* have to bring out the universal behavior of the whole world, and land ourselves in antinomical reasonings. *Would-be* objects are but appearances, fleeting shadows constantly put together by the exercise of synthesis, so that they could be extracted from the opaque contexts[6]—the behaviors—to which they belong, and pronouns and quantifiers be used to refer to them, only to attain failure every time, and more than that: *necessary* failure. Contexts will remain ineluctably opaque; quantifying in will remain forever foreclosed.

Nowadays an alternative way of making the same point, distant in rhetoric if not in content from Kant's transcendental idealism, is far more popular; and it is important to understand that the point being made is the same. Reality, it is said (in the framework I have been supplying here, the interaction of behaviors—I will be using similar words, like "world" o "universe" with the same meaning), is chaotic: factors that are negligible to the point of not being perceived at all may end up having catastrophic aftermaths. *That is*: since no universal behavior is (consistently) conceivable, there is always going to be an *outside* to any behavior—something the behavior does not contemplate. It is not a matter of scale: it does not have to be a butterfly flapping its wings; it could be a galaxy grumbling. (It is a matter of scale within transcendental realism, where one foolishly presumes that the universe is given, and hence if something is not contemplated it must be very *very* small.) What does matter is that, for any behavior, there is something foreign to its scope. And, once the behavior is expanded to include that foreign material (once the material is "perceived"), it can bring about changes of momentous proportions—typically, it is so included *because* it *has*, startlingly, brought about momentous changes. A chaotic world is where one lives after appreciating the significance of Kantian antinomies (and, as such, I can and will use that terminology freely in what follows).

As for the other, obvious candidates to the role of objects, hinted at above when I spoke of *mathematical* precision, Kant claims that eternally immutable numbers and geometrical figures do not exist if not insofar as they are incarnate in the appearances of the empirical world, the only one to which he attributes any kind of reality. But we need not follow him there: it is enough to point out that the supposedly absolute steadiness of their behaviors only works if we artificially restrict their range. To be sure, 7 will always be $3 + 4$, no exception; but 7 can also become my favorite number, and then I can change my mind about it and turn to favor 10 instead; and how could we justify the claim that such changes are irrelevant to 7—are not changes *of 7*—if not by arbitrarily setting the stage so that it confirms our bias?[7]

[6] For this notion of (referential) opacity see Quine (1956, 1961).

[7] As the example makes clear, mathematical entities are not necessarily alien to spacetime, though an effort can be made—and has constantly been made—to *keep* them that way: to *abstract* them from any spatiotemporal range where they occur and only consider their non-spatiotemporal behaviors. Which, as I said, are not going to be my concern here.

All subjects, therefore, are open to a certain amount of play, and play can bring about an evolution of their behavior, though not always a positive one—no providential account of evolution is entertained here. In an appropriately expanded form, a billiard ball's behavior when hit by a stick can show a relatively steady course of routine runs around the table interspersed with awkward ones when the ball is given an unusual spin, then a variety of attempts at governing the spins, and then a new series of runs that have the spins successfully incorporated. Or it can show repeated fiascos in governing the spins followed by hitting that has become exceptionally awkward, as if dominated by frustration and rage. What is not positive from one point of view, however, can be so from a different one. If we further expand the behavior above to include two billiard balls hit by different individuals A and B, we can see A perform decently, if not outstandingly, by hitting in a straightforward, far-from-fancy manner, until he tries harder, gets nowhere, becomes irked, and starts losing badly, game after game. His behavior's evolution has been to B's advantage, not to his own. When play brings about an evolution of a behavior, we say that the behavior (and its subject) has *learned*—and, again, there are not only positive cases of learning: self-destructiveness can be learned too (psychologists speak of a learned helplessness).

Though there can be no objects, hence no subjects whose behavior is absolutely steady, some subjects (and behaviors) make room for only a minimal amount of play, and others for a large amount. The former allow for sizable regularity and predictability, and for the currency of scientific disciplines, where everything works according to schedule (the future resembles the past—though it is not *identical* with the past: there are always little dissonances that go unnoticed, or neglected) until it no longer does (the butterfly flapping its wings becomes a hurricane, or the grumbling galaxy starts sending radiation that could cook small planet Earth). By setting up a narrow enough context (the range of operations for what Kant calls the understanding), those disciplines can manage reasonably well for ample periods of time; and, the more they do, the more they generate a misleading, perilous sense of security and even further resistance to play, so that, when the irregular and unpredictable happens, responses are even more crystallized into stereotypes and unable to match the novel challenges. Since crises (hurricanes, or threatening radiation) manifest themselves quickly, new responses cannot be improvised: there must be a vast repertory of them thoroughly developed (learned) before they could be relevant, in a disinterested way (this is the purposiveness without an end Kant talks about in the third *Critique*[8]). It can happen, and certainly ought to happen if subjects and behaviors are not to collapse (dissolve into rain), within those (other?) subjects and behaviors that make room for maximum play, and hence may have resources available to face what is new and unpredicted, for themselves and for others. (After trying out spins for a while, and becoming proficient at them, an individual can spread his learning to friends and colleagues, making them better prepared for when they meet a hustler.)

Most often, these latter subjects and behaviors fall within what are socially labeled the arts or the humanities, though the boundaries are porous. A "scientist" like

[8] See Kant (2002, 111).

Einstein played boldly (and irresponsibly, and dangerously) with matter, energy, and light; a "scientist" like Charles Davenport was adamantly assured of his, and his "science's" steadiness while heralding eugenics with words like "People don't have heated discussions on the multiplication tables. They will not dispute quantitative findings in any science."[9] On the other hand, if painting can give us new eyes and music new ears, and the visions seen by the ones or the resonances heard by the others can conjure up new worlds, or at least new corners of one, then the rigmarole of commercial exhibitions or stuffy concerts only encourages more trite conformity. If combining words in songs or paragraphs can provide new grammars for sentiment and action, then factory-producing bestsellers for passengers on long-distance travel or insomniacs on long-lasting nights only reinforces quotidian confabulation, with all of its humdrum tropes. If thinking of a single substance with infinitely many attributes and modes, or of monads shut from the outside and programmed in preestablished harmony with each other can guide imagination along fruitful avenues, then the plethora of departments, journals, conferences, and book series in something presumptuously labeled "philosophy" only seals vacuous, repetitive chatter within airless crypts.

Play, then, and the learning it contributes, is the engine of survival, for behaviors and subjects, for neither will be able, however much they expand, to keep themselves forever steady, and hence must be equipped to take a dive, now and then, into the stream of change. But, as we continue to take lessons from our examples, play itself is exposed to multiple dangers, which threaten its nature and dull its potential efficacy. On the one hand, there is all the stuff, and more of the same, I hinted at above: a travesty of play that, under the deceptive cover of luster and frills, delivers anything but an evasion from the familiar and the conventional. We can see how it operates in an expanded view: a subject (say, late consumer capitalism) that draws gigantic dividends from silly, monotonous practices will engage its victims in "playing" the lottery, or the slot machines, or the fans of professional sports, or the "friends" connected by a chatty social network, or the "connoisseurs" of the latest "artistic" fad; and these gloomy rites will infest (like parasites, and with like consequences) their free time, making it no longer free, subjugating it to the same sorry charades that populate their "busy" time. (To maximize its safety, this subject will also, while actual work by the victims is less and less needed, add to their busyness with *make-work*.) At the end of the day, they will have accumulated no new resources from such insipid fare fraudulently disguised as play; indeed, as with the "sciences" that get more brittle the more complacent they get, they will have lost some of their agility and become stiffer and more easily breakable.

On the other hand, play is pushed to further levels of audacity by competition. In order to win my next game of pool, I may spend hours practicing unheard-of spins and surprise even myself with the intricacy of the nuances I can achieve moving the ball. My opponent may do the same; and, in a behavior that includes both of us, we may then see unfathomable change come forward for both of us, enabling us to react skillfully to all possible fluctuations of events around the table (though neither,

[9] Quoted in Okrent (2019, 257).

following upon this arms race, ends up winning more). The same will be true in any other competing environment: Bjorn Borg, who once reigned virtually unrivaled over tennis, would have trouble today mastering any of the top one hundred players. But this is itself a problem: one more danger play must face. For, along such lines, play gives way to *games*: competitive episodes (in some of which one may well compete with oneself: solitaire is also a game) rigidly identified by *rules* where *play*fulness is *ruled out* to the extent that it does not subscribe to them (one plays *at* the rules but not *with* them)—the more so when we go from a single episode to a collection of episodes or an abstract definition of them: from playing a single game with rules that were just made up this once to playing within an official game structure that becomes, to put it in Wittgensteinian terms, a form of life. One can, by turning a game (of the latter sort) into a job or an obsession, become plenty agile within it, but only to the detriment of any other form of agility—for all intents and purposes, he will be a one-trick pony. And, if demands were put on him requiring moves not comprised in that one trick (if the world were not reducible to a pool table or a tennis court, and its trials could not be handled by carefully engineered spins or surgically administered backhand shots), he will be at a loss. Daniel Dennett (2006) speculates on how drastically different, and just as unnervingly complex, a game would be that coincided with chess except that the king could move two cases instead of one, and how deeply puzzled by its intricate logic people might be—whatever their proficiency at regular chess. Food for thought.

The risk of falling into unilateral, and self-defeating, specialization could be addressed by expanding the behavior, and the subject, under consideration. At the very beginning of an organism's life, its totipotent cells can develop into cells of any kind, even of organs external to the organism like the placenta or the yolk sac: they are the most versatile. (Italo Calvino at times voiced his fascination with the beginnings of novels, and beginnings in general, where a story, or a life, or the world, could still go any number of different ways,[10] and eventually wrote a book made—among other things—of ten such beginnings.[11]) With the organism's development, we go from totipotent cells to pluripotent ones, then to multipotent, oligopotent, and unipotent ones, progressively losing that versatility. But in a human organism, if liver cells can only perform liver jobs and muscle cells muscle jobs, the organism as a whole can perform both of those jobs and many more; so most of what is lost at the individual cell's level is retained at the level of the organism. Analogously, a society could be constituted of individual humans (or ants, or bees) each of whom is only good for one job, whereas the society as a whole could be good (by utilizing individuals) for

[10] "Every time the beginning is this moment of separation from the multiplicity of possibles: for the narrator it is the distancing from oneself of the multiplicity of possible stories, so as to be able to isolate and tell the single story he has decided to tell tonight. […] The beginning is the literary locus par excellence" (Calvino 1995, 735; translation mine). "So is what matters, in everything, only the beginning, the moment when all energy is tensed, when only the future exists?" (Calvino (1971, 14).

[11] Calvino (1981). There we read about the dream of "a book that is only an *incipit*, that maintains for its whole duration the potentiality of the beginning, the expectation still not focused on an object," 177.

all of them, whether they be games or more "serious" occupations. And this society could even show a high amount of playfulness though all of its individual members are obsessively dedicated to a single game. (Plato's ideal republic does not get there: its individual members are each locked into a single job, but the whole society is also thought to be radically—and unrealistically—steady.)

Which brings up another crucial point. Though play is inescapable, it may emerge only when a behavior is expanded up to a certain level. It may take cells enslaved to the dullest conduct for the organism that includes them to be able to play freely; in order to see the play, then, we must rise at least to that level. (And, of course, the cells are always open to start playing on their own, to become cancerous, and to make the organism unable to properly function, not to mention play.) So, if play is the engine of survival, then another, for me more basic, sense of competition gets in the picture: competition as to *who*, or *what*, is to survive; *who*, or *what*, will exploit play as an instrument of his, or her, or its own survival. An organism's play will be of no advantage to the individual cells, which will be used up and discarded without a chance to meet novel tasks with a novel attitude; individual consumers will not profit from enriched perspectives and fresh tactics as consumer capitalism forces them to engorge on trash at the mall. And politics—the art or science of negotiating among conflicting claims and powers—thus gets entrenched in the purportedly serene field of metaphysics, the science of being qua being: what survives, therefore ultimately what is, is what wins, however feebly and temporarily, the battle of what plays and learns as opposed to what is forced into an aberrant steadiness, for however long one is able to keep the latter in that aberrant condition.

Before we get to artificial, digital cases of this kind of competition, let us look at a few "natural" ones (while conveniently forgetting that artifice is but nature by other means). A river can only playfully meander because of the unsurpassable resistance that individual drops of its water encounter against particles of rock: the drops do nothing other than hit rock and be sidetracked; the river pleasantly traipses. Mountain faces playfully show straight, angular shapes because of the irresistible push their molecules were subjected to (no play there) in being lifted up from the Earth's crust, and of the further pressure put on them by erosion processes: the molecules have no choice (and sometimes are downright lost); the mountain face shines proudly in the light and contends for the golden apple with its peers. The Sun, the Earth, and the Moon dance in a chaotic three-body system while each chunk of them can only go this way or that—nothing chancy or undecided there.

Step now to the plate of bloodier, more violent confrontations. Slaves are forced into iterative, mind-numbing (a revealing term that!—I will return to it) activities so that their masters can enjoy amiable pastimes and improve their agility in various quarters (if you were Aristotle, in the pursuit of wisdom); within the subject "slavery" the masters' play is conditional upon the slaves' steadiness (until the slaves revolt). Workers on the assembly line meet the same fate, and so correspondingly do *their* masters: Fritz Lang's *Metropolis* gives a bleak portrait of the situation; Charlie Chaplin's *Modern Times* a more hilarious (but just as sad) one. And today the new slaves at Amazon and Uber are going over the same terrain. Native Americans are jailed into reservations, with strongly reduced opportunities for play, so that white people can roam freely in the prairies. Palestinians are massacred (who does not survive cannot play—but they will come back, as we are about to see!) so that Israelis can make merry in Westernized, "democratic" fashions.

And there are the animals: the "other" ones, since humans grudgingly accept the fact that they are animals too only to immediately raise an impassable barrier to block out any other species of the same genus.[12] They cannot speak, they cannot think, they cannot reason; they are stupid brutes. And a god made in the image of humans has put them all to the humans' service. The "progressive" linguist Noam Chomsky has been quoted as saying: "It's about as likely that an ape will prove to have a language ability as there is an island somewhere with a species of flightless birds waiting for humans to teach them to fly."[13] Which is patently false: many animals have their own language, which we are beginning to understand, and there is no reason why they should take exams in human language to pass muster in linguistic ability. Animals think and reason; they have aesthetic experiences and artifacts, and moral codes. What is *not* false is that they have been subdued to the worst form of slavery, to work the fields or just to be slaughtered, and in the death camps of intensive, industrial farming they are indeed stupid brutes; but that only as a result of the treatment they suffer. It is an understatement that there they have no room for play; and, with no play, they will enjoy no development of any faculties, linguistic or otherwise, no learning—which is true of humans, too, when they suffer the same treatment.[14]

Animals were regarded by Descartes as machines, incapable of feelings and in general of anything like a human experience. In the *Histoire de la philosophy cartésienne*, by Francisque Bouillier,[15] we read that once, as the Cartesian philosopher Nicolas Malebranche was entering the Oratoire Saint-Honoré in the company of Bernard de Fontenelle, and the pregnant house dog approached Malebranche to show affection, he kicked it. Fontenelle protested, and Malebranche replied: "Don't you know that it feels nothing?" Which means: it behaves *as if* it felt pain, because it is programmed that way, but it is incapable of real pain. Chomsky is a rationalist like Descartes and Malebranche: one of his early books was called *Cartesian Linguistics: A Chapter in the History of Rationalist Thought*.[16] And the sorry state of contemporary philosophy of mind is a witness to this resurgence of Cartesian dualism, with its private, ineffable *qualia*, philosophical zombies, and what it is like to be a bat. All just-so stories based on a cavalier interpretation of the task of conceiving or imagining: as Dennett (1995) aptly put it, "when philosophers claim that zombies are

[12] In "L'uomo e/è la scimmia" (in my (2010, 37–55) I revealed, and countered, the "transcendental creationism" to which many progressive, even radical, thinkers cling in spite of their adopting "empirical evolutionism." Chomsky, to be brought up presently, is one of them.

[13] See, for example, Chomsky (2007/2008).

[14] In addition to these cases of play simply being denied to some for the benefit of others, there are also cases, one of which will be mentioned shortly, in which play is allowed to some but they derive no learning from it (whereas others do). One way we can understand the progress of artificial intelligence, as I point out at the very end, is that it is moving from a situation in which humans are often in the second condition to one in which they might be stuck in the first.

[15] Bouillier (1868, 155).

[16] Chomsky (1966).

conceivable, they invariably under-estimate the task of conception (or imagination), and end up imagining something that violates their own definition" (322). A point that goes back to Wittgenstein (1953): "'But in a fairy tale the pot too can see and hear!' (Certainly; but it *can* also talk.) 'But the fairy tale only invents what is not the case: it does not talk *nonsense*.'—It is not as simple as that. Is it false or nonsensical to say that a pot talks? Have we a clear picture of the circumstances in which we should say of a pot that it talked?" (97).

I do not intend to engage here in a criticism of Descartes, Chomsky, or the philosophy of mind; I am not writing that kind of a book. My book is constructive, not critical. But, in order to construct, one must first clear the ground of debris—not speculate about why the ground should be cleared, just clear it. So I will make it explicit that I reject Cartesian dualism; that I believe a mind metaphysically separate from bodies and existing independently in its own sphere should go (and get lost) the same way as the God of *Genesis*, the blowing of holy spirit into mud, and all the rest of that drivel. Keeping things separate (minds from bodies as much as blacks from whites) is no less a political operation than is easing survival and being; but, however strong a force is exercised to preserve this political state, the return of the repressed is forever looming. And mind will enter my theorizing later on; not, however, as such a childishly conceived, arrogantly imposed character.[17]

This chapter can fittingly end with a first reference to computer networks, which also brings me back to the first intuition I had of the view I am presenting here. It was in 1990, before the official date of birth of the World Wide Web; what I was looking at was bulletin boards, and was seeing interesting phenomena take place there. Individual (human) users were letting their defenses down, as if they were in a private space, though they were not[18]: they were communicating with others—other individual (human) users. They were making offbeat suggestions, and picking them up from one another, and developing them; they were, in a word, playing. Not just with content: with emotions, too. "Flaming," that is, the venting of irritation and abuse, was a common occurrence. But they were not necessarily learning from their play: often, they were throwing content and emotions into the network with no intention or hope of getting anything back; they were behaving, I said then, "like King Midas' barber: digging a hole in the ground and screaming in it, only to then close it and bury all that the screams implied (or promised)."[19] Could something else be simultaneously playing as they did but, in contrast with them, be profiting from its play? Maybe the network itself? Could the individual (human) users be themselves used by a superior organism, a subject (a self—these two words will be interchangeable here) larger than themselves and unrecognized by them? Could

[17] An additional use of Cartesians (as well as of other authors, like Aristotle or Frege) to be made below is in setting up their views in opposition to my own. But that is intended as no criticism, as I have no time or patience to spare for that sort of thing here: it is rather the usual tactic of clarifying a view (in this case my own) by comparing it and contrasting it with others.

[18] The notion of privacy will be discussed in Chapter 6.

[19] My (1991, 523). Notice that in the mythological story there is a network too: a bed of reeds grows from the closed hole and it continues to whisper "King Midas has an ass's ears."

learning from play happen at that level, while the individual (human) users got no wiser for it?

I am stopping here for the moment; much more will be said about the matter in later chapters. What will be said is, however, nothing other than the articulation of this idea, now thirty-five years old and already articulated enough then, and in subsequent pieces,[20] that its structure could be seen clearly. Which it was not; so, to give it a chance to be seen now, I am going to work it out (I repeat) in painstaking detail.

References

Bencivenga, Ermanno. "The Electronic Self." In *Advances in Scientific Philosophy*, edited by Gerhard Schurz and Georg J. W. Dorn. Amsterdam: Rodopi, 1991, 517–528.

Bencivenga, Ermanno. *Giocare per forza. Critica della società del divertimento.* Milano: Mondadori, 1995. (1995a).

Bencivenga Ermanno. "On the Electronic Self Again: An Interview." *My Kantian Ways. Berkeley and Los Angeles*: University of California Press, 1995. (1995b), 23–32.

Bencivenga, Ermanno. *A Theory of Language and Mind*. Berkeley and Los Angeles: University of California Press, 1997.

Bencivenga, Ermanno. *La filosofia come strumento di liberazione*. Milano: Cortina, 2010.

Bencivenga, Ermanno. *Return from Exile: A Theory of Possibility*. Lanham MD: Lexington Books, 2013.

Bencivenga, Ermanno. *Meditazioni metafisiche. Contro Cartesio e in nome di una cagna gravida.* Napoli: Guida, 2023.

Bouillier, Francisque. *Histoire de la philosophie cartésienne*, vol. I. Paris: Ch. Delagrave, 1868.

Calvino, Italo. *The Watcher & Other Stories*, translated by William Weaver and Archibald Colquhoun. New York: Harcourt Brace, 1971.

Calvino, Italo. *If on a Winter's Night a Traveler*, translated by William Weaver. New York: Harcourt Brace, 1981.

Calvino, Italo. *Saggi*. Milano: Mondadori, 1995.

Chomsky, Noam. *Cartesian Linguistics: A Chapter in the History of Rationalist Thought*. New York: Harper & Row, 1966.

Chomsky, Noam. "On the Myth of Ape Language: Noam Chomsky Interviewed by Matt Aames Cucchiaro," Electronic Mail Correspondence 2007/2008. https://chomsky.info/2007____/.

Dennett, Daniel. "The Unimagined Preposterousness of Zombies." *Journal of Consciousness Studies* 2 (1995), 322–326.

Dennett, Daniel. "Higher-Order Truths about Chmess." *Topoi* 25 (2006), 39–41.

Kant, Immanuel. *Critique of the Power of Judgment*, translated by Paul Guyer and Eric Matthews. Cambridge UK: Cambridge University Press, 2002.

Lattimore, Richmond Alexander. *The Odyssey of Homer*. New York: Harper & Row, 1967.

Okrent, Daniel. *The Guarded Gate*. New York: Scribner, 2019.

Plato. *The Republic*, translated by Allan Bloom, second edition. New York: Basic Books, 1991.

Quine, Willard Van Orman. "Quantifiers and Propositional Attitudes." *Philosophical Review* 53 (1956), 177–187.

[20] See at least my (1995b); "La rete," in my (1995a, 149–186); (1997).

Quine, Willard Van Orman. "Reference and Modality." In *From a Logical Point of View*. New York: Harper & Row, 1961, 139–157.

Wittgenstein, Ludwig. *Philosophical Investigations*, translated by Gertrude Elizabeth Margaret Anscombe. New York: Macmillan, 1953.

Chapter 2
Intelligence

Abstract Intelligence is a single quality, that applies to behavior: a behavior (and its subject) shows intelligence to the extent that it overcomes difficulties in ways that are novel to it, though not necessarily novel in any absolute sense. Intelligence is manifested everywhere in the world: in plants, in animals of all kinds (not necessarily mammals, or even vertebrates), in transportation systems, and in machines. In all cases, intelligent behavior is behavior that has learned useful lessons from play. The case of plants is especially interesting, because it directs our attention to an intelligence manifested not by an individual tree but by a whole forest. The case of machines received useful focus when Turing devised his test for deciding whether a machine can think. That test went counter to millennia of an anthropomorphic understanding of intelligence, evident in Aristotle (most colorfully, and unfairly, in his many disparaging comments on the octopus) and recently revived by Searle with his Chinese-room argument. In fact, Turing already anticipated Searle's criticism by pointing out that machines had until then limited experience, and we had limited experience of them; all it takes for them to perform closer to humans is that they have as large an experience as we do.

Keywords Intelligence · Plants · Octopus · Turing test · Chinese room

In the last few decades there has been a lot of yakking about multiple intelligences, most of it traceable to Gardner (1983). As I do not intend to participate in the yakking (but I will offer a diagnosis for it below), I state unambiguously that I take intelligence to be a single quality, which I define as follows: a behavior (and its subject) shows intelligence to the extent that it overcomes difficulties in ways that are novel to it, though not necessarily novel in any absolute sense. The behavior does not have to be logical, linguistic, or even human; and in fact my first example of one will come from the world of insects—among the many unfortunate beings to which Chomskyans resolutely deny any sort of language ability.

I once had mice in the garage, and decided to use a glue trap for them. One morning I realized that a lizard (poor thing!) had been caught in it, and was dead. I did not dispose of the trap right away and several hours later, when I looked at it again, I was

E. Bencivenga, *The Philosophy of (Artificial) Intelligence*,
https://doi.org/10.1007/978-3-032-09347-9_2

treated to an astonishing sight. Ants had carried tiny specks of soil from a flowerbed some four feet away—a gigantic distance compared to their size (Irvine ants are quite small)—, had made a land bridge with them through the glue, had reached the lizard, and were proceeding to take it away in pieces. The trivial, catch-all notion of instinct does not begin to do justice to what had happened; it is the last refuge of the lazy and the coward in front of, indeed, an obvious display of intelligence. On the time scale of natural evolution, glue traps are something that happened a minute ago; it is even conceivable that ants, those particular ones or any others, had never before encountered a predicament just like the one they were facing then. But they addressed it creatively and successfully, and I can only surmise how they did it. They may have adapted to the new task moves that had proved sound elsewhere (crossing wet patches? fishing prey from small puddles?). They must have proceeded by trial and error (there were quite a few dead ants in the glue). Sure enough it worked.

My next example comes from Mancuso (2024), and is one of many he has brought out of plant intelligence. In that book's third chapter, "Planting the Underground," Mancuso writes of two New Zealand researchers, Sebastian Leuzinger and Martin Bader, who, "during an excursion in a rain forest north of Auckland, [...] discovered a stump of a kauri tree (*Agathis australis*) that had some unusual characteristics" (63). A stump is what is left of a dead tree; but that particular one was "undoubtedly active. Although it had no leaves or other organs capable of performing photosynthesis, its internal tissue was still living" (ibid.). How was it possible? "A tree without leaves, in fact, is not only deprived of its energy source—not being able to photosynthesize the sugars that allow it to live—but it is also virtually incapable of absorbing from the surrounding terrain the water it needs. The transpiration force of the leaves is the engine that allows the water to be absorbed from the ground and pumped into the tree. In short, without leaves, a tree not only is unable to photosynthesize but is also deprived of its water supply" (ibid.). For all intents and purposes, the stump appeared to be a true zombie (not an imaginary one, as in philosophy-of-mind mental experiments): a true living dead.

The resolution of the mystery was that the stump "received what it needed through its root system, which was directly connected with the root system of nearby trees thanks to a phenomenon known as root grafting" (ibid.). This case is interesting for me in two ways. First, it shows another admirable instance of intelligence at work, in solving the not-so-trivial problem of maintaining life in the absence of any individual resource. But, second (and differently from other brilliant instances of plant intelligence also described by Mancuso, as when plants emit substances that attract predators of the insects attacking them, or inject into predators other substances that turn them into cannibals), it also clearly shows that the proper point of view on it is not, really, the individual's at all. The relevant subject, here, is not the stump but the forest, or, as Mancuso calls it, the "vegetable community" (65). After applying sensors to the stump and the neighboring kauri trees, Leuzinger and Bader were able to ascertain "a close hydraulic relationship between the neighboring trees and the zombie stump" (68). So the intelligence displayed did not belong to the stump but to the community, which makes the situation analogous to the one hinted at above (and to be brought up again later) in which the relevant subject of an electronic exchange is not an individual user but a network. In Mancuso's words:

> The story told by these living dead stumps and the underground communications between distant trees is something so new and fascinating as to change our very conception of what a tree is. Regardless of the differences among species, trees have always been considered isolated individuals. Sure, not individuals in the animal sense of "indivisible" but indisputably single living organisms with needs and behaviors distinct from those of their fellow trees. In this respect, trees are no different from animals. But is this really true? Hundreds of research studies conducted over the last twenty years seem to be demonstrating a completely different reality. Not single trees but rather enormous connected communities, which through their root systems are able to exchange nutrients, water, and information. Extended communities that sometimes even include plants of different species and that depend for their survival more on cooperation than on competition. A true conceptual revolution whose consequences are not easy to predict. (66)

Descartes' tree of knowledge[1] is ready to be dumped then (if it was ever of any use), and with it the rigidly hierarchical structure he assigned to humans, and by implication to all of nature: a mind mysteriously communicating with the brain through the pineal gland, and through it utilizing the whole body as a machine (as well as, by implication, all of nature, since it, aside from minds, is only constituted of machines).

This hierarchical, vertical model is also present in Aristotle, though not without tension and conflict.[2] His official line is that, if understanding proceeds from principles to their logical consequences, and from universals to particulars, then the very shape of humans (or, rather, of men, since for him women are but mutilated males[3]) is a metaphor of this process: man's higher faculties (those germane to principles and universals) are on top, and contact with the dirty details of existence is to be found further down. Consistently with such an arrangement, he says in *Sense and Sensibilia*:

> [I]t is hearing that contributes most to the growth of intelligence. For rational discourse is a cause of instruction in virtue of its being audible, which it is, not in its own right, but incidentally; since it is composed of words, and each word is a symbol. Accordingly, of persons destitute from birth of either [...] [hearing or sight], the blind are more intelligent than the deaf and dumb.[4]

This characterization of intelligence fits well with the hierarchical model, where the best understanding is the most general, the kind expressed in conceptual thought and shared by rational discourse (that one must be able to hear—which is why, early in the *Metaphysics*, we are told that animals "incapable of hearing sounds [...] cannot be taught"[5]). But there is tension, I said—specifically, there is this other passage,

[1] See his *Principles of Philosophy*, in Descartes (1985/1991, vol. I, 186).

[2] The working out of this issue is taken from my (2008).

[3] "[T]he female is, as it were, a mutilated male." *Generation of Animals*, in Aristotle (1984, 1144 (737a)).

[4] Aristotle (1984, 694 (437a)).

[5] Ibid., 1552 (980a).

from *On the Soul* (similar ones can be found in *Sense and Sensibilia*,[6] *History of Animals*,[7] and *Parts of Animals*[8]):

> While in respect of all the other senses we fall below many species of animals, in respect of touch we far excel all other species in exactness of discrimination. That is why man is the most intelligent of all animals. This is confirmed by the fact that it is to differences in the organ of touch and to nothing else that the differences between man and man in respect of natural endowment are due; men whose flesh is hard are ill-endowed with intellect, men whose flesh is soft, well-endowed.[9] (So how about women, Ari?)

Quoting from myself now:

> What is going on here is that two conflicting models of intelligence are facing off. According to the official one, intelligence manifests itself in rigorous, clear-cut distinctions, such as can best be made from a distance with detached senses like sight and hearing, and, ultimately, by abstracting—separating out—universals from the confusion that is typical of all the senses. It is only by means of this separation that the eternal essences, hidden behind the colorful, delusive surface show of ordinary experience, are revealed. The second model of intelligence is less conspicuous but equally present in the Aristotelian corpus. According to that model, intelligence develops by manipulating things and environments, and in the process getting used to them, habituating oneself to them, learning to be efficient with, and in, them, even finding pleasure in feeling this manipulation reach high levels of efficiency. When explaining how to educate an excellent man (that is, a man who is most a man), Aristotle describes in painstaking detail and with great sensitivity this second model, and you come away thinking that, despite what the *Metaphysics* says, it is not memory and hearing that constitute the crucial factors in learning—not unless you take learning to be parroting—but a much more direct and concrete being-with your surroundings, internalizing routines and reactions, building stamina and resistance in the process, so that by the time you are required to express the mental state of courage or temperance you will not find your body giving way and your good intentions collapsing into *akrasía*.[10]

As often happens when there is conflict, there is a hapless victim here: the octopus. No other natural kind receives as much abuse in the Philosopher's corpus as this one. In *History of Animals*, it is labeled "a stupid creature,",[11] and in *Parts of Animals* we learn that it has a brain only "by analogy"[12] (we will see that that is true to a point, but it is *no* weakness). It does not seem to have much fun, either: *Generation of Animals* informs us that, despite what fishermen say, "the insertion [...] of the tentacle of the male into the funnel of the female [...] is only for the sake of attachment" and no evidence of copulation.[13] *Parts of Animals* describes how nature makes an exception to its ordinary arrangements for one variety of this unfortunate creature, whose suckers are not "the most advantageous arrangement, but [...] [just a] consequence

---

[6] Ibid., 700 (441a).

[7] Ibid., 787 (494b).

[8] Ibid., 1028 (660a).

[9] Ibid., 670 (421a).

[10] My (2008, 369–370). *Akrasía*, and Aristotle's take on it, will be discussed in some detail in Chapter 7.

[11] Aristotle (1984, 968 (622a)).

[12] Ibid., 1016 (652b).

[13] Ibid., 1119 (720b).

of the special nature of their substance."[14] Misfortune continues after the creature has been killed and cooked. In one of the *Fragments* (from Plutarch), Aristotle says: "To show that some foods bring bad dreams and disturb our sleeping visions, they adduce beans and the head of the octopus, from which those who resort to divination through dreams are ordered to abstain."[15]

Why so much lack of sympathy? Because the octopus uncharitably uncovers the Philosopher's bluff. Whereas he, in the end, would want to subordinate concrete, manipulating intelligence to its abstract, mental variety, and claim that courage, temperance, and other virtues of character, admirable as they are, are only instrumental to the true vehicle of *eudaimonía*—the study and contemplation of eternal truths, to be done uniquely with that head which reaches for the sky—the octopus would have none of that: it will continue to lack any definite structure, any clear hierarchy, any up and down. "It is particularly hard to distinguish [the uterus] in the case of the octopus, so that it seems to be single [as opposed to double, which it otherwise always is], but the reason of this is that the bulk of the body is everywhere similar" (*Generation of Animals*[16]). Moreover, "[t]he cephalopoda [the class to which the octopus belongs, and one that is not especially popular in these texts either] are the only locomotive animals in which the distinction of upper and lower does not exist" (*Generation of Animals*[17]). In the cephalopoda, "the two extremities are brought together by a curve [...] [so that] the excrement is voided near the mouth" (*Parts of Animals*[18]). And, as we are generalizing to a wider scope of abuse, let us remark how the cephalopoda lay "imperfect" eggs (*Generation of Animals*[19]) and how, a few pages after stating that "nature makes nothing contrary to nature" (*Progression of Animals*[20]), Aristotle goes on to say, of molluscs (the even larger class that includes cephalopoda), that they "do move, but move in a manner contrary to nature. They are not moving things, but are moving if regarded as sedentary creatures and creatures attached by growth, sedentary if classed with progressing animals".[21] Even when it comes to the celebrated quote from the *Politics* that "man is by nature a social animal,"[22] the octopus gives it the lie by showing what true sociability, authentic friendliness, is: how it is consummated in touching and squeezing and feeling, how it demands the physical warmth of other animals, the physical texture of other skins, the physical embrace of other arms. Only to be severely chastised, once again, for

[14] Ibid., 1069 (685b).

[15] Ibid., 2448.

[16] Ibid., 1113–1114 (717a).

[17] Ibid., 1151 (741b).

[18] Ibid., 1068 (685a).

[19] Ibid., 1138 (733a).

[20] Ibid., 1105 (711a).

[21] Ibid., 1109 (714a).

[22] Ibid., 1987 (1253a). Here the translation offered is "man is a political animal"; but the emphasis of the passage is on how natural it is for humans to live in a community; therefore, if the State is by nature, so are all "earlier forms of society" (ibid., 1252b) and the State is but the culmination of them, hence the alternative translation "man is a social animal" has been adopted by many.

its stupidity: additional evidence of it is that the octopus "will approach a man's hand if it is lowered in the water" (*History of Animals*[23]) and that "[t]he female after parturition [...] becomes stupid; if tossed about by waves, it submits impassively; a man, if he dived, could catch it with the hand."[24]

But, despite all this bad press by "the Master [...] of those who know,"[25] the octopus is a marvel of intelligence. It has been proven to learn by trial and error, to solve puzzles and untie knots, to navigate mazes, to use tools and fit objects into openings, to recognize individual people and behave distinctively toward them. It is similar to trees (or forests) in that, though it does have a central brain (which trees do not), its intelligence is (as in trees) distributed through its entire body: there are twice as many nerve cells in its eight arms as in the brain, so that the arms can process information and act on their own. Peter Godfrey-Smith has claimed that cephalopoda represent an independent branch in the evolution of intelligence—independent from the one traveled by birds and mammals—and that studying them is "probably the closest we will come to meeting an intelligent alien."[26]

One could say that Aristotle's and Descartes' misunderstanding of the nature of intelligence is due to their anthropomorphizing it: trees and octopuses don't look like us, so they must be stupid, they must be mere machines. But no: though I will counter anthropomorphism throughout (and I will specifically attack issues of looking like us—of scale and shape—in Chapter 9), here I find it worth noting that the model upheld by both, and many others, does violence as much to humans as to anything else. It is not anthropomorphic but mentalistic (or psychic, or animistic, if you prefer to take your lead from *psyche* or *anima* rather than *mens*); it spurns and degrades the body—in line with a tradition that goes back to the very inaugural scene of Western philosophy: the one in which a man is being executed and expresses relief in finally finding himself (*that is*, his immaterial soul) liberated from the body's unwieldy and vicious burden.[27]

So let us get into the human realm, and do so by an indirect route, by discussing an example of intelligence that, though it involves humans, can be attributed to no individual member of the species, and to which no individual human name can be fastened. I take it from Konrad Lorenz's 1973 Nobel Lecture.[28] When the steam engine made it possible to carry much longer and heavier loads by railway than had ever been carried by horse- or mule-driven carriages, an Aristotelian-Cartesian exercise of intelligence would have amounted to people sitting around a table (or in armchairs by the fire) and deciding by theory and argument in what shape it was most rational that the resulting vehicle (the train) be built. But this is not at all what happened. Let us hear from Lorenz how the story unfolds:

---

[23] Ibid., 968 (622a).

[24] Ibid.

[25] Alighieri (1867, 15).

[26] Godfrey-Smith (2016, 9).

[27] See Plato (1977, 30–35 (79e-84b)).

[28] Lorenz further developed this and related examples in his (1978).

> The ancestral form of the horse-drawn coach stubbornly persists, despite the very considerable difficulties which it entails, such as the necessity of constructing a runningboard all along the train, on which the conductor had to climb along, from compartment to compartment, exposed to the inclemency of the weather and to the obvious danger of falling off. The advantages of the alternative solution of building a longitudinal corridor within the carriage are so obvious that they serve as a demonstration of the amazing power exerted by the factors tending to preserve historical features in defiance of expediency. (102)

In other words, coaches—the same ones that horses drew, here relabeled compartments—were tied to one another, and it was only after a long process of adaptation that the contemporary torpedo (or cigar, or suppository) shape was attained. In the passage above, Lorenz stresses the factors impeding the process—hampering an intelligent development—: the stubbornness of ancestral forms, the inertia of established habits.[29] For us, however, it is more important to point out the positive factors—those that in fact turned the tide. If they did not consist in abstract reasoning from principles to consequences, or from universals to particulars, what they did consist in was tinkering with details, attempting local adjustments, trying a few things here and there until some of them seemed to perform better. The trains I have seen in my lifetime, and those of earlier times I have seen pictures of, are all testimony to this gradual improvement, driven not by a clear image of the objective to reach but by a blind bumping into solutions for problems that are only then first seen and which just for that reason are *novel* solutions, collectively manifesting a historic unfolding of intelligence.

Moving now more directly into the activities of individual humans, mathematics has always been the most revered field of application of the vertical model of intelligence: there the body does not enter; every result is obtained by logical reasoning.[30] As mathematics is codified today in formal systems, where theorems follow from axioms, or former theorems, by the use of rules, one could conceive of its growth as purely mechanical: rules are used, in some specified order, on the last theorem proved, and that use brings about new theorems. And on and on. But that is not what people, or even mathematicians, revere about mathematics. I remember my algebra teacher (Cesarina Marchionna Tibiletti) voicing her praise of Lagrange's theorem, by which (for finite groups) the order (number of elements) of a subgroup is proved to be a divisor of the order of the group, in something like the following words: what is remarkable about this theorem is that it connects structures (groups, subgroups)

[29] One might add existing (and powerful) economic interests. As a legendary case (in both senses of exemplary and somewhat fictionalized), consider the Tucker 48, a highly innovative car designed by Preston Tucker (in 1948) that featured (among other things) a rear engine, fuel injection, disc brakes, four-wheel independent suspension, and seat belts but never quite made it to the market due, at least in part, to the hostility of the big three automakers (General Motors, Ford, and Chrysler) which did not intend to give up their predominance (and/or to be forced themselves into extensive, and expensive, innovation).

[30] Notice that, consistently with my disregarding non-spatiotemporal behaviors, here I am discussing mathematics not as an abstract structure but as an activity of humans—not mathematics per se, if you will, but *mathematicians*, their work, and their values. And, even though what they deal with is supposedly bodiless, their activities are not: they perform such activities, and act out their values, in space and time.

defined in the most abstract terms, with no reference to numbers, with a basic notion (divisor) of elementary arithmetic—the one practiced by kids in elementary school. Mathematicians themselves sense that what is valuable in their discipline is the bringing together of odd couples (or larger aggregates): that the perfunctory spitting out of one logical consequence after the other is not (or should not be) what mathematics, and mathematical intelligence, are about. The most creative among them are those who come up with the most surprising such associations, and that includes the one to whom we owe (by chance, as fate would have it) the very existence of computers and computer networks.

For Alan Turing, a computing machine was a metaphor, which allowed him to address a problem (the decision problem for first-order logic) that, on the surface, had nothing to do with machines or computing by following an original, circuitous tack (it was his wont to devise original, efficient approaches to tricky problems). The later contingencies of World War II, and of the sleuthing that went with it, made his playful image into a colossal reality, and then the playfulness materialized into one or more Colossus[31] won that war—or, at least, the battle of the Atlantic, decisive for the Normandy landings and for the Red Army not getting all the way to Paris. So for individual humans, too, even those situated (like him) at the very apex of intelligent behavior, intelligence is no other than what I defined it to be: it may shine within rational pursuits, but what it *must* do, both a necessary and a sufficient condition for acknowledging its presence, is that, with all types of pursuits, it throw new angles on them, it reshape and retool them, it answer the questions posed by them by changing their nature, by turning them into different questions. If you think of it the wrong (Aristotelian, Cartesian) way, you will not acknowledge it outside rational pursuits and, feeling at least a family resemblance there, you will be forced (this is the diagnosis I promised) into talk of multiple intelligences. If someone improvises music deftly, she will have musical intelligence; if she can creatively arrange her space, she will have visual intelligence; if she can move her body and other objects skillfully and artfully, she will have kinesthetic intelligence. A true miraculous multiplication of loaves and fishes, based on simple narrow vision.

The introduction of Turing just now is apposite, since his (1950) provides two precious inputs for continuing our discussion. The first one has to do with what is called there "the imitation game" and has come to be popularly known as the Turing test. In it, a human interrogator is to judge a natural-language conversation between a human and a machine, trying to tell which is which. (To avoid being influenced by tone of voice or other environmental clues, the interrogator only ever sees printouts of the exchanges.) The machine passes the test if the interrogator cannot identify it as such: if he cannot tell the human and the machine apart.

A key feature of Turing's method is that he begins by "propos[ing] to consider the question 'Can machines think?'" (433) but then, finding that question hard to handle, replaces it with another question "which is closely related to it and is expressed in

[31] Turing contributed to the design of a number of computers by that name, which successfully decripted the Lorenz cipher and thus made it possible for the British to read high-level German army messages during World War II. For details, see Hodges (1983).

relatively unambiguous words" (ibid.)—that is to say: he replaces it with his test. It was another case of his common practice (to which I alluded above) of replacing something that looked unmanageable with something else that permitted viable approaches; in this case, it is apparent that he replaces a highly abstract question with one that concerns behavior, thus showing himself a clear source of inspiration for my own work. (Not the only one; aside from Kant and Wittgenstein, I would mention at least a book almost contemporaneous to the Turing article: Ryle (2000), originally published in 1949—other sources will be brought up later.) But the later resurgence of Cartesianism found itself in equally clear opposition to any such suggestion, and in 1980 John Searle, who got to Descartes through misunderstanding Husserl and, under the circumstances, was merely regurgitating Leibniz in the *Monadology*,[32] had his own test to recommend[33]:

> Suppose that I'm locked in a room and given a large batch of Chinese writing. Suppose furthermore (as is indeed the case) that I know no Chinese, either written or spoken, and that I'm not even confident that I could recognize Chinese writing as Chinese writing distinct from, say, Japanese writing or meaningless squiggles. [...] Now suppose further that after this first batch of Chinese writing I am given a second batch of Chinese script together with a set of rules for correlating the second batch with the first batch. The rules are in English, and I understand these rules as well as any other native speaker of English. They enable me to correlate one set of formal symbols with another set of formal symbols, and all that "formal" means here is that I can identify the symbols entirely by their shapes. Now suppose also that I am given a third batch of Chinese symbols together with some instructions, again in English, that enable me to correlate elements of this third batch with the first two batches, and these rules instruct me how to give back certain Chinese symbols with certain sorts of shapes in response to certain sorts of shapes given me in the third batch. Unknown to me, the people who are giving me all of these symbols call the first batch "a script," they call the second batch a "story," and they call the third batch "questions." Furthermore, they call the symbols I give them back in response to the third batch "answers to the questions," and the set of rules in English that they gave me, they call "the program." [...] Suppose also that after a while I get so good at following the instructions for manipulating the Chinese symbols and the programmers get so good at writing the programs that from the external point of view—that is, from the point of view of somebody outside the room in which I am locked—my answers to the questions are absolutely indistinguishable from those of native Chinese speakers. Nobody just looking at my answers can tell that I don't speak a word of Chinese. (417–418)

Searle's conclusion is (as was Leibniz's) that drawing conclusions from my behavior is irremediably deceptive: the reality of the situation is that, no matter what my "answers" are, I "haven't the faintest idea what [...] [they] mean" (418). Therefore, I do not understand what I am doing, as a computer's wouldn't whose behavior I am mimicking, and I, or the computer, however good our performance

[32] "Moreover, we must confess that the *perception*, and what depends on it, are *inexplicable in terms of mechanical reasons*, that is, through shapes and motions. If we imagine that there is a machine whose structure makes it think, sense, and have perceptions, we could conceive it enlarged, keeping the same proportions, so that we could enter into it, as one enters into a mill. Assuming that, when inspecting its interior, we will only find parts that push one another, and we will never find anything to explain a perception. And so, we should seek perception in the simple substance and not in the composite or in the machine" (Leibniz [1991], 70).

[33] See Searle (1980).

on the Turing test, cannot be said to provide a positive answer to Turing's *original* question: we would not be thinking.

The Chinese-room argument is one of a family of maneuvers that try to deal with challenges posed by computers by changing the subject. Faced by the undeniable similarity of some human and computer behavior, supporters of an irreducible distinction between the two[34] will say that, yes, the behaviors are similar, *however*: computers have no feelings, or are not conscious, or cannot tell right from wrong, or (in Searle's pseudo-Husserlian view) are lacking in intentionality. And all such worries have a single, shaky ground: as long as computers are developed within a restricted range, and employed in performing restricted tasks, their *whole* behavior will exhibit only a restricted similarity to *the whole* of human behavior; as soon as that range and those tasks are expanded, they will come to resemble human behavior more and more. I will spell out the intriguing variations of this point below, so I will cut it short here, limiting myself to noting that Turing (1950) was perfectly aware of how shaky this class of "arguments" is. First, he groups them all together:

> *Arguments from Various Disabilities.* These arguments take the form, "I grant you that you can make machines do all things you mentioned but you will never be able to make one to do X." (447)

And then he undercuts them all:

> No support is usually offered for these statements. I believe they are mostly founded on the principle of scientific induction. A man has seen thousands of machines in his lifetime. From what he sees he draws a number of general conclusions. They are ugly, each is designed for a very limited purpose, when required for a minutely different purpose they are useless, the variety of behaviour of any one of them is very small, etc., etc. Naturally he concludes that these are necessary properties of machines in general. [...] These applications of the principle are of course largely unconscious. When a burnt child fears the fire and shows that he fears it by avoiding it, I should say that he was applying scientific induction. [...] The works and customs of mankind do not seem to be very suitable material to which to apply scientific induction. A very large part of space-time must be investigated, if reliable results are to be obtained. Otherwise we may (as most English children do) decide that everybody speaks English, and that it is silly to learn French. (447–448)[35]

The second invaluable input Turing provides has to do with the kind of programming that could be done with a computer. His speculation is both revolutionary and visionary:

---

[34] Remember how establishing an irreducible distinction between mind (or soul) and body was one of two main goals of Descartes' *Meditations on First Philosophy*, built into their very subtitle. See Descartes (1985/1991, vol. II, 12).

[35] Searle's paper referred to above is part of a panel discussion in which a first piece by him is followed by several comments by others and then by the author's response. And the initial piece itself is constructed dialectically, with a first version of the argument followed by the entertaining of various criticisms and by consequent refinements of that first version. At one point he considers Turing's (anticipatory) devastating criticism, when he discusses "The Robot Reply" (420), where the instructions provided to the experimental subject are inserted into a robot that moves about in the world and has a number of experiences there, thus expanding the range of what it can do. But he sees no substantial difference between this case and the original one, and as such is appropriately chastised by Dennett (him again!) in his comments, when he says: "Searle does not dwell on how vast a difference this modification makes to what we are being asked to imagine" (429).

> Instead of trying to produce a programme to simulate the adult mind, why not rather try to produce one which simulates the child's? If this were then subjected to an appropriate course of education one would obtain the adult brain. Presumably the child-brain is something like a note-book as one buys from the stationers. Rather little mechanism, and lots of blank sheets. [...] Our hope is that there is so little mechanism in the child-brain that something like it can be easily programmed. The amount of work in the education we can assume, as a first approximation, to be the same as for the human child. (456)

This idea "of a learning machine", says Turing, "may appear paradoxical to some readers. How can the rules of operation of the machine change? They should describe completely how the machine will react whatever its history might be, whatever changes it might undergo. The rules are thus quite time-invariant" (458). His response is that there can be different levels of rules, and different kinds of them. The rules that can change may be more local, context-dependent ones, situated within a general framework that stays time-invariant; and the most general (time-invariant) ones may be (at least in part) rules specifying how to change when meeting changing stimuli (they can be analogous to the open programs used by natural evolution and depicted by Konrad Lorenz in [1978] and elsewhere).

> For at every stage when one is using a logical system, there is a very large number of alternative steps, any of which one is permitted to apply, so far as obedience to the rules of the logical system is concerned. These choices make the difference between a brilliant and a footling reasoner [remember what I said above about creative mathematicians], not the difference between a sound and a fallacious one. Propositions leading to imperatives of this kind might be "When Socrates is mentioned, use the syllogism in Barbara" or "If one method has been proved quicker than another method, do not use the slower method." Some of these may "given by authority," but others may be produced by the machine itself, *e.g.* by scientific induction. (458)

A scientific induction that can be faulty, but can also use its own faults to correct and improve upon its earlier outcomes.

As I already announced more than once, a lot of what was introduced in this chapter will be enlarged and deepened in what follows. Now it is time to bring the chapter to a close by a brief recap of what these various instances of intelligent behavior have taught us. Intelligence is intimately related to play, hence to trying something, *essaying* it if you prefer a fancier word,[36] for the hell of it, without any utilitarian goal. Most of the tries are failures, as most biological mutations are monstrosities, and consequently fall by the wayside; but some may provide, as I said at the beginning, novel ways of resolving difficulties that threaten the efficiency, or even the existence, of a behavior; so the behavior (in an expanded form) will incorporate them, proving itself to have learned from them. For this to happen, the behavior's subject does not have to be a human: it can be an animal, a tree (or a forest), a transportation system, a weather configuration, a colony of ants or bacteria, a cancer. Or a computer (network). The first step in coping realistically with current AI is to dispel the anthropomorphic daze that blocks our vision.

---

[36] In the sixteenth century, Michel de Montaigne used the corresponding French word, that means attempting, experimenting, to label what he was doing in his work. And I like to mention that the views expounded here first coalesced for me in the course of writing my (1990) on him and were first applied by me to computers (as I mentioned above) in the same year of its publication.

References

Alighieri, Dante. *Divine Comedy*, translated by Henry Wadsworth Longfellow. Boston: Houghton, Mifflin, and Company, 1867.

Aristotle. *The Complete Works of Aristotle*, edited by Jonathan Barnes. Princeton NJ: Princeton University Press, 1984.

Bencivenga, Ermanno. *The Discipline of Subjectivity: An Essay on Montaigne*. Princeton: Princeton University Press, 1990.

Bencivenga, Ermanno. "The Trouble with Cephalopoda." *Common Knowledge* 14 (2008), 365–373. (2008)

Descartes, René. *The Philosophical Writings of Descartes*, in three volumes, translated by John Cottingham, Robert Stoothoff, Dugald Murdoch, and Anthony Kenny. Cambridge UK: Cambridge University Press, 1985/1991.

Gardner, Howard. *Frames of Mind*. New York: Basic Books, 1983.

Godfrey-Smith, Peter. *The Octopus, the Sea, and the Deep Origins of Consciousness*. Glasgow: William Collins, 2016.

Hodges, Andrew. *Alan Turing: The Enigma*. New York: Simon & Schuster, 1983.

Leibniz, Gottfried Wilhelm. *Discourse on Metaphysics and Other Essays*, translated by Daniel Garber and Roger Ariew. Indianapolis IN: Hackett, 1991.

Lorenz, Konrad. "Analogy as a Source of Knowledge," the 1973 Nobel Lecture. At https://www.nobelprize.org/uploads/2018/06/lorenz-lecture.pdf.

Lorenz, Konrad. *Behind the Mirror: A Search for a Natural History of Human Knowledge*, translated by Ronald Taylor. New York: Harper, 1978.

Mancuso, Stefano. *Planting Our World*, translated by Gregory Conti. New York: Other Press, 2024.

Plato. *Phaedo*, translated by Georges Maximilien Antoine Grube. Indianapolis, IN: Hackett, 1977.

Ryle, Gilbert. *The Concept of Mind*. London: Penguin, 2000.

Searle, John Rogers. "Minds, Brains, and Programs." *The Behavioral and Brain Sciences* 3 (1980), 417–457.

Turing, Alan Mathison. "Computing Machinery and Intelligence." *Mind* 59 (1950), 433–460.

Chapter 3
Soul

Abstract I espouse not empirical but transcendental behaviorism. It is silly to think that all there is to the world of empirical psychology, or the world at large, is behavior. It is true, rather, that in thinking about the world of empirical psychology, or indeed the world at large, and in talking about it, we should build our logical space of concepts and our vocabulary of words starting with the concept of behavior, and every other concept we use should be either dependent on, or defined in terms of, that of behavior. Within this thinking and talking, there is room for the traditional notion of soul: not as an autonomous entity (let alone an immaterial, immortal one) but as a quality of behaviors and subjects, which are soulful (or, even better, soulfully) insofar as they are playful. In agreement with classical figures of the Italian philosophical tradition like Giordano Bruno and Tommaso Campanella, I believe that everything in the world is alive and soulful, having a drive to survival (Bruno's "desio di conservarsi"). And I also agree with Bruno in claiming that there is no survival without self-transcendence, whose archetypes are the Heraclitean fire evoked by Hegel in picturing his concept, and the love that, says Bruno, "converts the loved thing into the lover in the same way as fire, which of all elements is the most active, is powerful in converting all those other simple and composite things into itself."

Keywords Behaviorism · Soul · Bruno · Campanella · Fire · Love

In Searle's 1980 exchange with his critics, the word "behaviorism" occurs on both sides, largely as a term of abuse:

> [I]n much of AI there is a residual behaviorism or operationalism. [...] I believe that if AI workers totally repudiated behaviorism and operationalism much of the confusion between simulation and duplication would be eliminated. (423; Searle speaking)

> Any plausibility attaching to the idea that a Schank machine [the specific target of Searle's Chinese-room argument] thinks would seem to derive from a crude Turing test version of behaviorism that is anathema to most who view cognition as formal symbol manipulation. (426; Ned Block speaking)

E. Bencivenga, *The Philosophy of (Artificial) Intelligence*,
https://doi.org/10.1007/978-3-032-09347-9_3

And there is some variously-striped, uniformly-fuzzy concern about distinguishing acceptable from unacceptable varieties of behaviorism:

> While the crude version of behaviorism is refuted by well-known arguments, there is a more sophisticated version that avoids them; however, it can be refuted using an example akin to the one Searle uses against Schank. (426; Ned Block speaking)

> But now we might have worked out exactly which kinds of operationalism and behaviorism to be ashamed of and which not. (446; Richard Rorty speaking)

Because the very first notion I introduced in this book was that of behavior, and everything else was built on its basis, it may be in order to pause for a moment and explain in nonfuzzy, straightforward terms how I stand with respect to behaviorism. Here I find it, once again, profitable to take my lead from Kant, and more precisely from his distinction between empirical and transcendental idealism. The former, which he rejects, is the silly view (advanced by Berkeley, and a constant threat for Descartes and sundry rationalists) that the world of experience is literally constituted of ideas—whereas such an empirical realist as himself sees it constituted of objects, however deficient (phenomenal) their being might be. The latter, which he champions, is the view that, in *thinking* about the world of experience and *talking* about it, we should build our logical space of concepts and our vocabulary of words starting (not so much with the concept of an idea, which for him has a knottier function, but) with the concept of a *representation* (*Vorstellung*), and every other concept we use should be either dependent on, or defined in terms of, that of a representation. The concept of an object is dependent on that of a representation (an object is what a representation represents) and, given the irreducibly phenomenal nature of objects, will never become wholly independent of it; the concept of a cognition is defined by putting a number of (categorial) conditions on that of a representation.[1]

Analogously, I distinguish between empirical and transcendental behaviorism (while noting that similar distinctions have been attempted by many, as suggested above when speaking of acceptable and unacceptable forms of it, or kinds of it to be or not to be ashamed of, but without a lucid recognition of where exactly the distinction should be located—the closest anyone got to stating it clearly was Ryle [2000]). The former, which I reject, is the silly view that all there is to the world of empirical psychology—or even the world at large—is behavior. The latter, which I champion, is the view that, in thinking about the world of empirical psychology, or indeed the world at large, and in talking about it, we should build our logical space of concepts and our vocabulary of words starting with the concept of behavior, and every other concept we use should be either dependent on, or defined in terms of, that of behavior. The concept of a subject is dependent on that of behavior: a subject is what behaves. The concept of a steady behavior is defined as that of a behavior which does not change; the concept of an object is defined as that of a subject which, no matter how far its behavior expanded, would always have a further expansion showing it to be steady.

---

[1] This interpretation of transcendental idealism was first formulated in my (1987).

A less highfalutin, more down-to-earth account of my stance would be to say that one can legitimately use any concept or word one wants as long as one can show what difference it makes in behavioral terms (which implies that it makes some): how behavior is different when the concept or word is properly applied and when it is not. Then the concept of being inspired by God would not be legitimate, since *any* behavior can equally be God-inspired or not: there are no behavioral traits that would allow for discrimination between the two kinds. (In this, I find myself in agreement with Turing when he says, still in the 1950 article and while answering a theological objection to the value of his test, "I am not very impressed with theological arguments whatever they may be used to support," 443.) But for other concepts that are often appropriated by theology it does not look (to me) as if the situation is so bad. Case in point: the soul (or mind, or *anima*). I noted already, in bringing up the death of Socrates, that I distance myself from the tradition seeing the soul as a substance metaphysically independent of the body, which can (and, for Christians and Platonists alike, will) lead a life after the body rots away (for Platonists, but not for Christians, even before the body comes to be). And yet, many of the intuitions which have long found themselves at home in that tradition can be recovered, and made good (if heterodox) sense of, in the picture I am drawing.

An object, we saw, is a nonentity: there aren't any; no subject can be permanently locked into steady behavior. But many subjects, we also saw, are to some extent objects, or object*ive*: their behavior is, to some extent, steady, and as such is studied by scientific disciplines, which can (based on how objective they are) predict where they will go and what they will do next. Occasionally, they will violate predictions: they will change and they will play—and to such an extent will no longer be objects (though they can return to that temporary status, when they stop playing and may have absorbed what learning the play induced). For what they will be during their playful phase I use the traditional terms *souls* and *minds* (interchangeable here); I will say that subjects are souls or minds to the extent that they are playful. And, on these terms, I now make two remarks: one to connect them (or, rather, their use) to what I said about behaviorism; the other to bring out another general feature of my overall approach (and language).

The first remark is that, as I use them but not as the tradition uses them, souls (and minds) can always be cashed out in behavioral terms (hence their use is legitimate). A traditionally-inclined person X may think that a human who looks, speaks, and moves just like the Terminator (Arnold Schwarzenegger, maybe?) differs from the Terminator *only* in the important (for X) regard that he has, but the Terminator has not, an immaterial and immortal soul; therefore, having or not having a soul would make no difference for X (and for the tradition to which he is inclined) in the behavior of the two—or of anything else. But, for soul to enter my discourse, there must be features of the behavior to which it applies that are ostensibly different from behaviors to which it does not apply. There is no soul to a steady behavior; there is to one that is playful; and a playful behavior is ostensibly not steady.

The second remark has to do with parts of speech. "Soul," "mind," "behavior," and "subject" are nouns and, in an ordinary language that is structured in essential agreement with an Aristotelian framework, they are supposed to refer to substances, or things, or objects (all the same here)—whereas, approximately, adjectives are supposed to refer to qualities and verbs to actions. But my framework is not Aristotelian; even more dramatically, there are in it no objects to refer to. Therefore, by explicating it in ordinary language I am doing violence to both—the framework and the language—and coming up with locutions and sentences that must feel unnatural to a speaker of that language, while they don't do full justice to those who see (see what? things? the same issue raises its head) my way. I could of course invent another language, or jargon, but I don't think that "nothing nothings"[2] is a model to imitate; I prefer to hassle with a language that resists me, and when I discuss language later it will become clear why. In this vein, I should alert the reader to the caution exercised in saying "there is (no) soul," rather than "there is (not) *a* soul": the latter would make soul an individual, countable member of a kind, assimilating it totally, in speech if not in fact, to an object; the former hovers uncertainly in an area where soul could be a predicate, designating a *feature*, a mode, a way, but not an object—though, damn it!, "feature," "mode," and "way" are nouns, too. A further step in that direction would be to reject the noun "soul" altogether in favor of the adjective "soulful," or perhaps the adverb "soulfully," and hence to replace "this behavior has soul" with "this behavior is soulful (or, less idiomatically, soulfully)." Which, however, would still leave untouched the problem that a behavior is not an object either, though it is referred to by a noun, nor is it what behaves.

Making do with the available lexicon and syntax, then, we could say that what is objective in a subject is also what is *dead* in it: that in it the future resembles the past means that there is really no future, that to speak of what we continue by inertia (!) to call the future is really to speak about the past, about what is settled and mummified into a fixed shape. *And* we could say that play, or soul (true to the latter's Latinate variant), *animates* a subject, brings *life* to it, if life is what oozes novelty, what breaks molds, what augurs surprise. Words like "body" or "matter," on the other hand, could be used to designate what in a subject is stiff and needs to be animated: the objective, dead aspect of it into which soul injects vitality.[3] Empirical death would then seem to be the reduction of a subject to pure body or pure matter, the vanishing from it of all soul; though, as there are no objects, we must admit that that, too, is but a temporary state—that the body (or matter) can again be animated (the past can be revived) in a properly expanded form of the same subject/behavior, where what was supposed to be forever done with becomes the "inspiration" for new excitement and gleeful commotion.

As a subject does not have to be human, or even animate in the Aristotelian sense, in order to play, every subject can have soul (in fact, at some point, *will* have it) and can be animated in my sense: can be alive. Which is where the antagonistic character of my use of language, remarked above, begins to show in full force. In accounting

[2] This refers to Heidegger's sentence "Das Nichts selbst nichtet" ("The nothing itself nothings [or noths, or nihilates]"), to be found in his 1929 lecture "What is Metaphysics?" (see it in his (1993, 93–110), where the fated sentence is on 103, with the translation "nihilates")—a lecture chastised by Carnap (1959).

[3] Though it should be noted that most often, when I use "body" in what follows, I am referring to an animated body, not to dead meat.

for it, it will be apt to unearth some other sources of *my* inspiration: some majestic figures of the Italian philosophical tradition.

Both Giordano Bruno (1548–1600) and Tommaso Campanella (1568–1639) defended the view that everything in the universe is alive. Here is how Bruno states it in his dialogue *Cause, Principle, and Unity* (the two interlocutors are Dicson and Teofilo—the latter, whose name literally means "lover of God," is Bruno's spokesman; the former's name is a reminder that the dialogue was written during a Bruno sojourn in England):

> *Dics.* I seem to be hearing something very unusual. You hold perhaps that not only the form of the universe, but all forms of natural objects whatever are soul.
>
> *Teo.* Yes.
>
> *Dics.* All things are then animated?
>
> *Teo.* Yes.
>
> *Dics.* But who will agree with you on this?
>
> *Teo.* Who with reason can disprove it?
>
> *Dics.* Common sense tells us that not all things are alive.
>
> *Teo.* Common sense is not the truest sense.[4]

The most elementary manifestation of life—whatever scope one takes life to have—is a drive to self-preservation and, true to form, Bruno claims that "every thing tends with all its forces to the preservation of itself".[5] In more detail:

> Nature, just as it gave being to every species, and gave to every thing the desire to preserve itself in its current state, also installed in every thing a certain spirit, or if one prefers sense, by which through an internal requirement it would know and flee the most serious dangers. This can be seen [...] in all things where, even if they seem deficient or dead, still resides a spirit that tries with all its forces to preserve the current condition. It happens with drops of a liquid which, when they are about to fall, form a sphere to avoid breaking off [...]. It can also be found in pieces of straw or wood thrown into the fire, and in tissues and membranes, which startle and withdraw to somehow avoid their own destruction. This sense is in every thing and is a form of life, even though, abiding to common usage, we do not call it animal.[6]

It makes one sad to think of Bruno's own tissues and membranes startling and withdrawing as he was being burnt at the stake in Campo de' Fiori, Rome, for asserting bold philosophical tenets like the ones above, on February 17, 1600 (a holy year, and the most massive participation of pilgrims ever seen there needed entertainment on a winter day, especially a Thursday at the very beginning of Lent). But then he will hearten us by pointing out:

> [M]any and innumerable beings live not only within us but also in all composite things [each subject is a locus of encounter for multifarious changing, expanding, contracting, playful behaviors]; and when we see something which is said to die, we must not believe that that

[4] Bruno (1964, 85–86).

[5] Bruno (2000, 646; translation mine).

[6] *De magia*, 406, to be accessed digitally at La Biblioteca Ideale di Giordano Bruno, Istituto Nazionale di Studi per il Rinascimento; translation mine.

> thing dies but rather that it changes and terminates its accidental composition and unity, since the things which we see incurring death always remain immortal.[7]

Therefore, "[t]he rational soul is not afraid of death; at times, indeed, it tends to it spontaneously, it spontaneously goes toward it."[8]

Campanella agrees, and adds valuable hints on the equally democratic distribution of sensibility, and the dangers of anthropomorphism:

> It is just as stupid to deny sensibility to things, because they have no eyes, or mouth, or ears, as it is to deny motion to the wind because it has no legs, or eating to fire because it has no teeth, or seeing to those living in the country because they have no window to look out from, or to the eagle because it has no eyeglasses.[9]

Even the air senses: "through light it sees, through motions it hears, through vapors it smells; through tenderness it tastes; and through pressure and heat and cold it touches; it feels pain, and pleasure, and, without organs, all of it senses."[10]

Talk of something being *preserved* in its current *state* or *condition*, however, can conjure up the wrong picture of what these people are after (once again, language must be fought against!). For there is nothing static to the drive summoned here: no attempt at keeping faithful to a stable configuration. Self- "preservation" is understood, rather, as self-transcendence: as a mighty impetus to extend oneself over one's other, to incorporate it in a larger version of oneself, and hence to *modify* one's status or condition, so that it be *not* kept the same. X will not (continue to) live if it does not (continue to) change. The archetype of this dynamic form of being is the very fire that consumed Bruno's tissues and membranes:

> Best evidence of it [of this irresistible tendency to grow] is fire, which now I show you enclosed in a small spark: if the stuff that makes up wood nourishes it, will it not grow to infinity and persist forever?[11]

And love works much like fire:

> [L]ove converts the loved thing into the lover in the same way as fire, which of all elements is the most active, is powerful in converting all those other simple and composite things into itself.[12]

There follows that loveless, changeless, non-transformative (cowardly!) life (the life lived by his judges!) is not even life, and that "no one is satisfied of his state, except someone who is dumb and foolish."[13]

Campanella's emphasis, once more, is on sensibility: he rejects the Aristotelian priority of sight and hearing, which receive information from afar without changing

---

[7] Bruno (1995, 157).

[8] Bruno (2000, 421; translation mine). When he was sentenced, Bruno told his judges that it was they who were afraid; they were the seat of incapacitating, sterilizing fear.

[9] Campanella (2003, 54; translation mine).

[10] Ibid. 146; translation mine.

[11] Bruno (2000, 459; translation mine).

[12] Bruno (2007, 534; translation mine).

[13] Ibid. 543; translation mine.

the object sensed or themselves, and assigns the highest value to (philosophically unpopular) taste—a sense that feels things most distinctly and informatively in the process of destroying them and amalgamating them into a changing, growing stage of oneself.

> *Sapientia* [wisdom] [...] comes from *sapor* [taste], and from taste that feels things as they are. This sense gives wisdom its name more than any other, since it not only perceives the external and strong natures of things but also the intrinsic, hidden, and weak ones. Touch knows in wine the cold that from the outside seizes wine and holds back its forces; but taste feels both the external cold and, through tasting it, the internal, hidden heat, which is the intrinsic heat arising when what is ingested gets broken down, or it flows penetrating into the pores of the tongue, hitting the spirit of its nerves, and revealing itself to them. [...] Therefore only this sense becomes intrinsic to the object, and the whole object to the sense; and wise is the one who knows things from the inside and not only superficially by the intellect.[14]

The wisdom philosophers search for, and the best kind of knowledge, are not to be attained by gaining the imperturbability of the Aristotelian *phrónimos*, peacefully at one with himself, but by participating in an unremitting transformation, in a melting of food into the fabric of one's own tissues; and Campanella further shows his resonance with Bruno when he claims that "[t]he human soul shows its infinity and immortality in loving and desiring, and whatever it might have acquired, be that even the whole world, it is not satisfied."[15] A soul, by the way, that is not infinitely plastic like Aristotle's, lacking its own form and ready to assume any other, to then lose it and return to virginity: a *corporeal* soul whose every contact with the outside leaves in it not the same kind of mark it would leave in a stone, but a mark nonetheless—every cognition the soul acquires shows up as a scar.

> [F]orms do not remain in the soul as in the stone, otherwise the soul could not receive new ones, but would always actually feel the same thing and would never forget it [...]: they vanish as if in the air; there remains, however, a mark, like a scar.[16]

For this vision which I share, that something is alive means that it plays, that it has soul. And, to reach the vision's pinnacle, only what is alive, and plays and has soul, *is*, has being; what is dead, to the extent that it is dead, *is not*. But Kant, who guides so much of my thinking, would seem to believe quite the opposite: the categorial conditions I mentioned above, that make for cognitions and ultimately should make for objects ("let us once try whether we do not get farther with the problems of metaphysics by assuming that the objects must conform to our cognition"[17] is how he formulates the "experiment"[18] known as his Copernican revolution), demand regularity, identifiability in spacetime, a distinction between substances and accidents, firm causal connections that define the very arrow of time—all sorts of features that impertinent play will call in question and throw into maddening chaos. What then? Is it time to part ways with Kant?

---

[14] Campanella (1994, 75; translation mine).

[15] Ibid. 289; translation mine.

[16] Ibid., 251; translation mine.

[17] Kant (1998, 110 (Bxvi)).

[18] Ibid.

Not quite. It is time, rather, to recognize that (at least) two conflicting paradigms are in play (!) concerning the notion of being: that metaphysics (as already mentioned) is a terrain of political struggle. On one side to the confrontation, being is what it was for Bruno and Campanella: a fiery cauldron where something can only "preserve" itself by turning into something "else," where what ordinarily goes under the name of preservation would be turning into a corpse (provisionally, as everything is provisional here). Coming two centuries after those Renaissance philosophers, astrologers, and magicians, Hegel gives it a good summary where he writes, of his concept (which for him, at an early dialectical stage, is all there is):

> The concept [...] is the simple negative into which all determination, all that is existent and individual sinks. [...] The concept is just the constant change of Heraclitus, the movement, the causticity, which nothing can resist. Thus the concept which finds itself, finds itself as the absolute power before which everything vanishes; and thereby all things, all existence, everything held to be secure, is now made fleeting.[19]

And Luigi Pirandello, one more century later, also gives it a good shot:

> Supposing, in a word, that there were no such thing as this death which fills us with such terror, that death should prove to be not the extinction of life but a gust of wind, merely, which blows out the light in our lantern, extinguishes this dolorous, painful, terrifying sense of life we have—terrifying, because it is limited, narrowed, fenced in by the circle of fictitious darkness that begins just where the light from our lantern stops. We think of ourselves as fireflies astray in this darkness, desperately casting about us tiny circles of radiance which are powerless to dispel the gloom, and which are, as it were, our prisons cutting us off from the universal, the eternal life to which we shall some day be allowed to return. Whereas, in point of fact, we are part of that greater life already, and always shall be, but henceforth without, let us hope, that feeling of exile and exclusion which torments us so.[20]

On the other side, being (real) is consistency, and no one puts it better than Descartes when he finally resolves the doubt based on the dream argument:

> If, while I am awake, anyone were suddenly to appear to me and then disappear immediately, as happens in sleep, so that I could not see where he had come from or where he had gone to, it would not be unreasonable for me to judge that he was a ghost, or a vision created in my brain, rather than a real man. But when I distinctly see where things come from and where and when they come to me, and when I can connect my perceptions of them with the whole of the rest of my life without a break, then I am quite certain that when I encounter these things I am not asleep but awake.[21]

Of these two sides, I favor the first: being, in the position I have been expounding, is more than dynamic—it is dynamism itself. And regularity is nonbeing. But it is also part of this position that any given situation is a compromise between the two: it is being to a point and nonbeing, also, to a point. The two, in fact, are always joined. In Sartre's words:

---

[19] Hegel (1995, vol. I, 352; translation modified).

[20] Pirandello (1923, 206–207).

[21] Descartes (1985/1991, vol. II, 61–62).

> [I]f nothingness can be given, it is neither before nor after being, nor in a general way outside being. Nothing lies coiled in the heart of being—like a worm.[22]

This compromise is what Kant is delivering: his "objects" are not like Charles Davenport's multiplication tables: an absolutely steady shore from which to expel the unfortunate (irredeemably unfortunate) specimens of inferior races.[23] They are not the equally absolute certainty of the Cartesian *cogito*: the rock solid base on which to erect an unwavering new edifice of knowledge.[24] They are ghostly entities, precariously held together by synthesis and ready to disappear as soon as a gust of wind blows out the light of the lantern that provides them what little radiance they can afford. The categories that articulate such providing also spell out their fragility: their incapacity to provide answers out of context, their dependence on a context, their inevitable exploding into incoherence when the context is denied. That the shakiness can be ignored for a long time, generating what I called a misleading, perilous sense of security, is irrelevant from a metaphysical viewpoint: no matter how durable the transcendental illusion, "objects" are but phenomena, and hence no objects.

As I argued in a variety of places,[25] this was for Kant part of his enterprise of "deny[ing] **knowledge** in order to make room for **faith**"[26]; more perspicuously, to curb classical metaphysical pretenses in order to defend the plausibility of ethics—from this (most revealing) angle, his entire critical corpus can be seen as the strenuous construction of a tenable metaphysics of morals. Here ethics will remain out of sight; so what I get from him is a weakening of metaphysical claims, including, most relevantly, his establishing in the Paralogisms that rational psychology (the "philosophy of mind" of Descartes, fellow rationalists, and their contemporary sidekicks) is a futile quest—there is for it no object of study; the mind that is supposed to be that object is none at all, in fact it contradicts the conditions of objectivity that grant empirical objects what ghostly status they have. In empirical terms, it is literally *nothing*—as Kant unfolds the various senses of this word in the closing section of the Transcendental Analytic, it is an *ens rationis*: "the object of a concept to which no intuition [...] can be given [...], and which must therefore not be counted among the possibilities."[27]

But there is a struggle, I said; and struggles are often fought on enemy land. So I will note again that "mind" is a noun, that for the enemy a name refers to an object, and that using it this way, this for me meaningless way, is a way of stressing in the

22 Sartre (1992, 56). Notice, however, that in Sartre being and nothing are reversed from the way I see them: his nothing is what I would call being, and vice versa.

23 Davenport's demented "science" was instrumental in setting drastic quotas for immigration into the U. S. from Asia and from Eastern and Southern Europe—which stayed in place for some forty years.

24 I will come back to this, and to how delusive it is, when I discuss *cogitare* (that is, thinking) in Chapter 6.

25 For a most detailed statement of this line of argument, see my (2007).

26 Kant (1998, 117 (Bxxx)).

27 Ibid., 382 (A290 B347).

enemy's face, and in his terms, the dignity of the position I defend, and he fights against. Quoting myself:

> "There is" no mind. And yet to claim that there is to make an important political statement. One that ought not, ever, to be "made true"—or its whole point would be lost. It must continue to be uttered foolishly, in defiance of the "facts," straining their "meaning"; it must keep on working as an extraneous body, rejected by the organism, carrying infection, breeding pus. If the mind did indeed become a thing, if "there were" a mind as "there are" tables and chairs, if a final reconciliation between the two were to obtain [...], we might as well forget about the mind.[28]

To pick up a thread left hanging above, it is this making of the mind into a thing, this stifling of the rebelliousness it urges by horrid sticky tactics, which amounts to mind-*numbing*: to the turning of audacity into rote.

References

Bencivenga, Ermanno. *Kant's Copernican Revolution.* New York: Oxford University Press, 1987.

Bencivenga, Ermanno. *A Theory of Language and Mind.* Berkeley and Los Angeles: University of California Press, 1997.

Bencivenga, Ermanno. *Ethics Vindicated: Kant's Transcendental Legitimation of Moral Discourse.* New York: Oxford University Press, 2007.

Bruno, Giordano. *Cause, Principle, and Unity*, translated by Jack Lindsay. New York: International Publishers, 1964.

Bruno, Giordano. *The Ash Wednesday Supper*, translated by Edward Alberic Gosselin and Lawrence S. Lerner. Toronto: University of Toronto Press, 1995.

Bruno, Giordano. *Opere latine*, edited by Carlo Monti. Torino: UTET, 2000.

Bruno, Giordano. *De gli eroici furori*, in *Opere italiane*, edited by Nuccio Ordine, vol. II. Torino: UTET, 2007.

Campanella, Tommaso. *Metafisica*, vol. I, edited by Paolo Ponzio. Bari: Levante, 1994.

Campanella, Tommaso. *Del senso delle cose e della Magia*, edited by Filiberto Walter Lupi. Soveria Mannelli CZ: Rubbettino, 2003.

Carnap, Rudolf. "The Elimination of Metaphysics through Logical Analysis of Language," translated by Arthur Pap in *Logical Positivism*, edited by Alfred Jules Ayer. Glencoe IL: The Free Press, 1959. (The original German appeared in *Erkenntnis* 2, 1932.)

Descartes, René. *The Philosophical Writings of Descartes*, in three volumes, translated by John Cottingham, Robert Stoothoff, Dugald Murdoch, and Anthony Kenny. Cambridge UK: Cambridge University Press, 1985/1991.

Hegel, Georg Wilhelm Friedrich. *Lectures on the History of Philosophy*, in three volumes, translated by Elizabeth Sanderson Haldane. Lincoln NE: University of Nebraska Press, 1995.

Heidegger, Martin. *Basic Writings*, revised and expanded edition by David Ferrell Krell. San Francisco: Harper, 1993.

Kant, Immanuel. *Critique of Pure Reason*, translated by Paul Guyer and Allen William Wood. Cambridge UK: Cambridge University Press, 1998.

Pirandello, Luigi. *The Late Mattia Pascal*, translated by Arthur Livingston. New York: Dutton, 1923.

Ryle, Gilbert. *The Concept of Mind.* London: Penguin, 2000.

[28] My (1997, 92–93).

Sartre, Jean-Paul. *Being and Nothingness: An Essay in Phenomenological Ontology*, translated by Hazel Estella Barnes. New York: Washington Square Press, 1992.

Searle, John Rogers. “Minds, Brains, and Programs.” *The Behavioral and Brain Sciences* 3 (1980), 417–457.

Turing, Alan Mathison. “Computing Machinery and Intelligence.” *Mind* 59 (1950), 433–460.

Chapter 4
Language

Abstract Play is not only useful (and pleasant); it is also risky; playing with fire can burn down the house. So safety measures are typically in place to protect players from risk. One is confining play within restricted, isolated areas: the playground, the laboratory, vacation time. Another is shifting play to substitute things: toy cars and guns instead of real ones. The most extensive and successful arena for substitute play is language, whose powerful and subtle articulation allows for mimicking complex situations and vicariously exploring them without encountering too many real dangers. Language is not, therefore, primarily or even importantly, a means of communication (as it is taken to be by a long tradition that goes from Aristotle to de Saussure and beyond) but, primarily and essentially, a field of play. "Description" of a reality in language, when it is worth anything, is the conjuring up of it; and the quality of the language is crucially involved in this operation, in that the reality will be believed, and people will want to live in it, depending on how engaging and fascinating the language is that "describes" it. A good example of this is Galileo, whom both Einstein and Stephen Hawking called the father of the new science, not because he was unique at his time as a brilliant mathematician, physicist, or astronomer (which he wasn't) or because he made no serious mistakes (which he did), but because his beautiful prose presented the "new science" as a thrilling, enchanting adventure and inspired readers to take part in it.

Keywords Risk · Safety · Language · Description · Galileo

It is time to look at some dividends of the Kantian compromise. So Bruno encountered fire in 1600, and his bodily tissues and membranes were destroyed. Say that he did not and they were not (at least then). It was, for all we know, a close call. He had recanted his heretical views and was about to be given a more lenient sentence when, unexpectedly and somewhat mysteriously, he pronounced a final, defiant withdrawal of his recantation and went on to meet a harsher fate. By then he was in his early fifties (his exact date of birth is unknown), and life expectancy in 1600, for someone who had survived infant mortality, was around seventy years of age. Disregarding the damage done to him, up to that point, by an erratic and embattled existence

E. Bencivenga, *The Philosophy of (Artificial) Intelligence*,
https://doi.org/10.1007/978-3-032-09347-9_4

(inclusive of torture), he could have counted, give or take, on another twenty years of maintaining those tissues and membranes in decent shape. Most likely, his best work was behind him, in what he had been writing, before his capture, at a frantic pace; it is hard to believe that, by kissing up to his judges, he could have expanded his behavior in a meaningful way. On the other hand, look at what expansion his tragic "death" has brought him (as a subject). His statue towers over the same Campo de' Fiori that witnessed it, sending a stern glance in the direction of the Vatican (it was erected in 1889, at a time when the Pope, after losing most of his temporal authority, regarded himself as a captive of the Italian State, by the freemason sculptor Ettore Ferrari). Though the enthusiasm surrounding him then has by now noticeably abated (the committee promoting that monument of his, which was founded in 1885, counted among its members Mikhail Bakunin, Ernst Haeckel, Victor Hugo, Heinrich Ibsen, and Herbert Spencer), his figure has remained a venerable symbol of bold standing up to tyranny. It's been quite a bargain: in exchange for a couple of extra decades of (probably grueling) physical presence on the planet, the subject Bruno has been granted extraordinary size and power.[1]

He's been an exception, though. Countless others have languished in the dungeons of despots, and have been cut to pieces by bloody executioners, with no notice of their agony reaching out and building momentum. It's a chaotic world, and whether a given behavior will expand or contract hangs mercilessly on the luck of the draw (on which butterfly flaps its wings); hence it makes as much sense as anything else not to go, always, all in—to play it more conservatively, closer to the chest. To bet on smaller expansions, to stay away from fire (from love too?—or maybe to *play* with love, without fully committing to it?); to keep the flickering lantern alight a little longer—as long as possible.

This is where categorial coherence looks promising. It only organizes small niches of cosmos, eventually to fall prey to Heraclitean fire; its predictions will hold up until, well, they no longer do; the more they hold, the worse hell might break loose when the castle of cards comes crashing down; but even this transitory respite may be likable, if the alternative is flesh and bones being roasted and incinerated in short order. So the best will be tried to abide by the outcome of synthesis, to highlight regularities, to parade causal links as evidence of solid control, to glorify "hard" sciences. And, if

[1] In light of this example, and of my general claim that subjects will return from the dead, you might wonder why I don't believe in an immortal soul (why I don't believe in a soul that is immaterial should be clear). And my answer is: First, a subject will go through periods of being totally inert—for all intents and purposes, dead, which justifies saying that it might return *from the dead*—hence it is not (and its soul is not) literally immortal. Second, the immortal soul of Plato, Christianity, and countless other philosophical and religious systems is taken to be *the very thing* that animated a particular body—indeed, its new destiny is totally determined by the vicissitudes that body had during its existence—whereas for me a subject returns to life by transcending its previous existence (just as, for Bruno, its previous existence itself was a constant act of transcendence). When Bruno comes back, it does so in his works and their readers and admirers, in his statue in Campo de' Fiori, in the influence he has on other thinkers; and any indiscretions that may have stained his previous existence are largely irrelevant. Therefore, by rejecting an immortal soul I also want to deny the most common understanding of that phrase.

no one breathes too loud, the wobbly structure might stay in place indefinitely: there is no telling which straw will break its back.

The laying down of that last straw might be postponed if the structure (the behavior) allows itself a certain amount of play, and of learning from it: play in a protected area, whence the spreading of subversion can be constrained and the risk of an avalanche shattering the environment be kept to a minimum. And whence, after a lot of cautious testing, some of the results can be applied outside the sanctuary. Think of experimentation in a laboratory: of how thick the walls are that separate it from the outside, how pervasive the protocols, how tightly sealed the space suits the operators wear. (And think of what happens when all this security, or even a minute portion of it, fails: when a virus engineered by gain-of-function research finds its way out of the lab before anyone can put a lid on it. True play with fire.) Assuming a behavior (in this case: the behavior of a society, or of a scientific community) can sustain a delicate balance between evoking frolic and avoiding doom, doom may have to wait a little longer as the behavior evolves new resources from its crucible. It is this delicate balance that Kant is recommending, what with his insistence on a priori necessities to uphold objectivity (even the objectivity of the self) on the one hand and his praise for creative scientists who "comprehended that reason has insight only into what it itself produces according to its own design; that it [...] must compel nature to answer its questions, rather than letting nature guide its movements by keeping reason, as it were, in leading strings"[2]—scientists like Galileo, Torricelli, *and Stahl*, to prove that, when you playfully pose questions of your own making to nature, the harvest may be prominent steps forward as much as embarrassing blind alleys.[3]

There are two main kinds of safety measures, of letting risky play occur while reducing the risk. One I just mentioned: keeping the play under wraps. Spatial wraps, as in a lab, or temporal wraps, as in vacations or holidays whose "violations" are not supposed to infect ordinary routines. Bakhtin's (1984) account of the carnivalesque gives good examples of the latter; "What happens in Vegas stays in Vegas" combines the two. If it's a child playing, locking the windows of a high-rise, securing the electric outlets, storing away the matches.

The other kind of measure I can also illustrate by reference to a child, and that is: letting it play with substitute implements, substitute characters, and substitute actions. The child would love to playfully approach a real car, and fiddle with all those intriguing buttons and levers and pedals, and turn its wheel this way and that to see how the vehicle reacts, but it will not be permitted to; what it will have available is a *toy* car, from which it will learn only a limited amount of what it takes to negotiate a real road—but, significantly, it will learn more, the more similar the toy is to reality. And it would love to be armed with a real gun as it duels in the Western sun with some other sharpshooter, or be cleverly managing a spaceship that travels close to the speed of light; but what it will get to do instead is impersonate those guys, play make-believe with them as it mimicks their countenances and feats (what countenances and

[2] Kant (1998), 109 (Bxiii).

[3] Stahl was a chief proponent of phlogiston theory, about which more will be said later.

feats it has seen sported by other, more professional impersonators) not in a desert or in space but on a playground.

It is an all-too-eloquent affair that what actors do on a stage be called *play*. For what they are doing there is what play is all about: expanding the range of their subjects, and vicariously of the audience's, by trying moves that, before they started rehearsing, might have been unfamiliar to them. But they are doing all of this at a one-step remove from everyday concerns, by bracketing the entire operation within a fictitious, fictional milieu. The ploy is far from foolproof; the fourth wall is constantly in danger of being broken, and a number of actors and directors have played (!) with that danger, with various levels of commitment and menace. *A Double Life*, a 1947 movie by George Cukor, shows an actor becoming so obsessed with the role of Othello as to commit murder in real life by his character's kiss of death (and, ironically, Ronald Colman, who played that actor, was so good an actor himself to earn an Oscar for his performance). Far more serious (no irony there) was the career of Jerzy Grotowski, who resisted throughout the notion of playacting and eventually, in his last semi-conventional production *Apocalypsis Cum Figuris*, asked actors to express their own personal tensions, misunderstandings, and mutual hostilities, and encouraged a mixing of actors and spectators, with both chatting casually, participating in scenes, and some of the spectators, when the time came for leaving, being asked to stay behind and collaborate with the team. Later reports by both groups of people bring out how much this experience had changed their understanding of not only the theater but also themselves. They had acquired knowledge in Campanella's sense: in the form of scars.

This threat which substitute play poses of infecting reality is what made Plato most worried about the theater, as he wanted (we saw) his republic to be absolutely steady and therefore decreed to ban all playful teasing with it:

> [I]f [...] [the guardians] do imitate, they must imitate what's appropriate to them from childhood: men who are courageous, moderate, holy, free, and everything of the sort; and what is slavish, or anything else shameful, they must neither do nor be clever at imitating, so that they won't get a taste for the being from its imitation. Or haven't you observed that imitations, if they are practiced continually from youth onwards, become established as habits and nature, in body and sounds and in thought?[4]

Imitation is thus both a viable channel for the learning induced by play, in that what begins as a simple "imitation game" (remember Turing) may "become established as habits and nature," and an auspicious (but not foolproof) mechanism for dodging dire consequences of it: as long as we are *only* imitating someone or something, we are not that yet—we are only a substitute for it. Depending on whether you see the glass half-empty or half-full, you will adopt opposite attitudes to it: anxious Platonists will want to enforce the most taxing checks, to make sure play does not hurt the stability of their system; affable admirers of Diderot or Montessori[5] might look favorably on the agility peacefully acquired through it. It is, once again, a balance of contrasting

---

[4] Plato (1991), 74 (395c).

[5] See Diderot (1883), Montessori (2002).

factors, and, while the balance is kept, a behavior can make a measure of progress without facing do-or-die, radically decisive circumstances.

By stretching colloquial English, we could say that, if X imitates Y or in general is a substitute for Y, it is also a *sign* for Y. A toy train would then be a sign for a real convoy; a child impersonating an astronaut would be a sign for the galactic wanderer; an actor playing Othello would be a sign for the moor of Venice. A collection of interrelated signs constitutes a *language*, which we can say to be *speaking of* the collection of what its signs substitute for. A toy train together with toy tracks and a toy station speaks of the corresponding conglomerate in real life; Thespians playing Othello, Desdemona, and Iago speak of the corresponding characters in Shakespeare's (and in a director's, and an audience's) imagination. It is by this route that we come to a new fundamental (and, for many, jarring) conclusion: language—any language—is but a way to practice the substitute strategy I just outlined, and to manage the rocky equilibrium it attempts between encouraging play and keeping it safe.

The story about language that is ingrained in our lore is both old and new, both a bestseller and a longseller. Language—we are told, with the kind of inexorable nodding approval that makes you suspect no one is listening any more—is a means of communication, of making information common across speakers and hearers (or writers and readers). More precisely, it is a tool for establishing contact between minds, so that what is at first only present in one mind can come to be shared by another. Here is Aristotle, in *De interpretatione*:

> Now spoken sounds are symbols of affections in the soul, and written marks symbols of spoken sounds. And just as written marks are not the same for all men, neither are spoken sounds. But what these are in the first place signs of—affections of the soul—are the same for all; and what these affections are likenesses of—actual things—are also the same.[6]

Forget the difference between English and Chinese; what matters in language is that, when it works, we get to talk about the same tables and chairs, and to do so via the same representations of tables and chairs (affections made by them) that are deposited in our minds (or souls). By the time Ferdinand de Saussure rehashes essentially the same scheme of things some twenty-two centuries later, he is ready to forget about tables and chairs too, and restrict himself to their mental representations (here relabeled concepts and images): linguistics is embedded, without residue, into psychology.

> [W]e must examine the individual act from which the speaking-circuit can be reconstructed. The act requires the presence of at least two persons; that is the minimum number necessary to complete the circuit. Suppose that two people, A and B, are conversing with each other [...]. Suppose that the opening of the circuit is in A's brain, where mental facts (concepts) are associated with representations of the linguistic sounds (sound-images) that are used for their expression. A given concept unlocks a corresponding sound-image in the brain; this purely *psychological* phenomenon is followed in turn by a *physiological* process: the brain transmits an impulse to the organs used in producing sounds. Then the sound waves travel from the mouth of A to the ear of B: a purely *physical* process. Next, the circuit continues in B, but the order is reversed: from the ear to the brain, the physiological transmission of the

[6] Aristotle (1984), 25 (16a).

sound-image; in the brain, the psychological association of the image with the corresponding concept.[7]

Despite talk of the brain, what we get here is mental communication, since the brain contains no concepts, and it is assumed, as in Aristotle, that the concepts (as well as the sound-images) are the same in the two minds; otherwise, I am not sure what A and B, in this scheme, would be saying to each other.

Descartes (if he could bring himself to believing in other minds, and resisting the suggestion that what he sees from his window is automata dressed in hats and coats[8]) would have approved, and may have suggested the pineal gland as the conduit between the brain and the mind. So it is not surprising that, within the eager contemporary Cartesian revival, yammer should have arisen and persisted about a *mentalese*, or language of thought: a language in which we are supposed to think (in which Descartes thinks the *cogito* to himself?) and which we are then supposed to encode, for purposes of communication, in the various spoken or written languages that we use for that purpose.[9]

I am not going to waste my time with this nonsense. I am only going to point out that, where people are seriously engaged in the business of deriving information from others' utterances—in interpreting texts, say, or unveiling psychological trouble—what might be going through the others' minds never enters the actual proceedings (however it might be vacuously mentioned). They do detective work. They construct theories about the reasons why someone spoke or wrote a certain way, and their high-level performance mirrors what anyone, more humbly, in fact does to figure out what a neighbor "means": advance and test conjectures about the profile of his whole comportment—verbal, written, physical, and social.[10] In doing justice to the intricacy of this work (even in the humble cases), a focus on how we process idiotic sentences like "Whales are mammals" (the opening example of the entry "The Language of Thought Hypothesis" in the *Stanford Encyclopedia of Philosophy*,[11] by Michael Rescorla, a suitably Harvard-educated specimen who is suitably doing time at UCLA) is a worthless distraction,[12] and will be given no further credit or attention. I did long ago, even longer than when I first tackled artificial intelligence, in my (1987), and I don't intend to reprise that chorus here. I will limit myself to a general, extreme comment which I also made before: if (spoken and written) language (forget mentalese) were a means of communication such as Aristotle and de Saussure (and their philosophy-of-mind minions) intend, it would have long ago become vestigial, as pelvic bones are in snakes and wings in flightless birds. What

[7] De Saussure (1959), 11–12.

[8] See Descartes (1985/1991), vol. II, 21.

[9] The yammer got started with Fodor (1975).

[10] This issue will come up again in Chapter 7, when dealing with intentions, and in Chapter 8 I will add important detail about it after introducing another useful perspective on language.

[11] See it at https://plato.stanford.edu/entries/language-thought/.

[12] Notice that the sentences used in such examples are uttered by no one, hence we are provided no context for interpreting them. Who knows what an actual person uttering "Whales are mammals" might mean by it? Perhaps she might be pointing to the stupidity of analytic philosophy of language.

I will substantively do, rather, is explain what language does *instead of* providing a means of communication.

Say that a child collected bottle caps for a while and now is organizing them as squadrons of Romans and Carthaginians to fight the battle of Cannae on the pavement of a courtyard. It wants to play with the historic course of action and outcome: offset Hannibal's pincer movement and have the legions emerge, if not unscathed, certainly victorious. But the material it is laboring with is highly inadequate: every cap is a substitute (a sign) for a soldier (or a company of them?) and, as long as they keep their orderly formation, they may put up a good showing; when they are to attack the enemy ranks, however, the masquerade brutally exposes its sketchiness. To show any purpose, the kid must move his "soldiers" one at a time, losing the unity of joint action; if, on the other hand, it pushes lots of caps together, then formation is lost and a classic case of military strategy turns into a detestable mess. Which also means, going back to what I said about the toy car, that little will be learned about fighting real battles: the resemblance between reality and its substitute is at a minimum, and so is its fallout.

It would be different if this were a videogame; but we need not go there—a more old-fashioned utensil will do. The child can tell himself (or others—around a fire, as fabulists once did?) a story, and then he will find in language all the litheness he was missing when dealing with caps, for language is finely *articulated* (it has finely delineated *artus*, limbs), and hence can do far more than provide blunt, rudimentary, all-or-nothing signs for complex references (one cap: one soldier). It also provides signs for the elements of those complexes, and for *their* elements, and for how they jostle or dance, and for the elaborate modes of their jostling and dancing. Verbal (and written) language, as opposed to the language we can "speak" with bottle caps, is a most pliable, supple repertory of substitutes, the most pliable and supple we are acquainted with, hence it is the most abundant, efficient setting to turn to for play that resorts to the substitute safety valve.

It all starts with the vocal cords, a folding of the throat that in grownups does not extend beyond the length of an inch but is able to utter a miraculously rich variety of sounds. A large part of that capability goes idle: when a Chinese student once voiced for me the four distinct inflections with which they pronounce any of their monosyllables (thus assigning each of them four different meanings), I (not being at this any better than Searle) could not tell them apart, hence also could not reproduce them. But what did not go dormant is rich enough to let us paint a world of boundless subtlety, and then play with it in the relative tranquillity of someone who only has to worry about swords and bombs and the plague as *objects of discourse*. Pace Chomsky, we are not unique in playing like this: birds and cetaceans do it too, and other species (bees, say) use linguistic tools different from sounds. But I have no qualms about admitting that, part as an effect of the genetic lottery, part as a result of training that went on for hundreds of generations, we may have the vastest stockpile available. If we are not (by a long shot!) the only users of languages, we may well be the most proficient ones.

Written language follows suit, with an additional complication. For spoken language is a continuous flow, whereas its written counterpart, increasingly in the last couple of millennia, has consisted of separate words, allowing for silent reading[13] but also creating problems for those who learn a second language, and have to find a way, after studying dictionaries and grammars, to cope with the torrent of undivided sounds thrown at them by natives.[14] Leaving that issue aside, the same combination of selective development and associated stunning bounty is to be found here: some of the symbols in languages other than ours, even if conveyed in some approximation to a Latin alphabet (a Latin alphabet, that is, supplemented with diacritic signs), we cannot *say*, nor does there seem to be any rhyme or reason to the words in which they occur (for a compelling case, I would cite a translation of one of my books into Finnish); yet what we are left with is enough to concoct an orgy of tales and to unleash endless substitute play.

One main problem with understanding the role of language is the harm that is constantly done by projecting a devious, mendacious ideology onto it, centered on a verb. Language is taken to *describe*, or alternatively and equivalenty to depict or portray, language-independent states of affairs. Wittgenstein (1961) is relevant here:

> 2.1. We picture facts to ourselves. (15).
>
> 2.12 A picture is a model of reality. (ibid.)
>
> 4.01. A proposition is a picture of reality. (37).
>
> 4.014. A gramophone record, the musical idea, the written notes, and the sound-waves, all stand to one another in the same internal relation of depicting that holds between language and the world. (39)

Language enters the proceedings, if we trust this propaganda, as always already subservient to facts, or reality, as providing a copy of it, most likely for the purpose of communicating information, and possibly instructions, to another. Wittgenstein again, this time from his (1953):

> Now think of the following use of language: I send someone shopping. I give him a slip marked "five red apples." He takes the slip to the shopkeeper, who opens the drawer marked "apples"; then he looks up the word "red" in a table and finds a colour sample opposite it; then he says the series of cardinal numbers—I assume that he knows them by heart—up to the word "five" and for each number he takes an apple of the same colour as the sample out of the drawer.— It is in this and similar ways that one operates with words. (2-3)[15]

This view can be extended without much fuss (it would seem) to cases in which facts, or reality, are not involved: where what we talk about is unreal characters like Achilles or unreal events like Gulliver's travels. For language then is still supposed to maintain its ancillary position with respect to the fake facts, or fake reality, of the speaker's or writer's imagination—and to be functional to conveying those contents to hearers or readers. It's so bad we should be using so many and so disparate conveyances for this sort of transmission, to the point where the transmission often

---

[13] See Saenger (1997).

[14] I have addressed this issue, in the context of a general discussion of continuity, in my (2025).

[15] Fortunately (for him and for us), in this and other works Wittgenstein developed a far more nuanced grasp of language; still, the passage above is a good representative of the caricature to which many have reduced it.

fails for lack of familiarity with the conveyance; perhaps one day we will be able to carry out such operations in mentalese!

I already complained on how idiotic it is to build a "philosophy of language" on test cases like "The cat is on the mat" or "Whales are mammals"; so I am not going to beat that dead horse again. What I will do, rather, is take the notion of describing—or, indeed, of depicting or portraying—at the opposite end of its use: not when it collapses into idiocy but when it shines with creativity. Look at Leonardo's *Mona Lisa*; is this a copy of anything? Is it supposed to make do for contact with its "original"? No: it is itself an original. There was probably a real woman early in the game, there was a project of portraying her, and there were intuitions in Leonardo's mind; but the painting transcends all of that and asserts itself as a world- and mind-independent archetype (which may be a reason why Leonard never delivered it to his client and kept it with himself until the end). We can call it an archetype of humanity, or beauty, or balance; but any of that would be beside the point. It is, first and foremost, an archetype *of itself*; after seeing it, we will not be able to look at, and think and speak of, humanity or beauty or balance or many other topics as we did earlier. Therefore, if language is depicting or portraying or describing, I ask that we take it to be doing those things—when it works well, when it does what it ought to do—in the same way as the *Mona Lisa*: that we take it to be inventing something that, barring its being thus put in *play*, would never have been. What language does when it is used to say "The cat is on the mat" or "Whales are mammals" I regard as a *pathology* of it, as would be a pathology for a human to be paraplegic: no longer to be able to dance. And trying to account for language starting from those pathological uses of it would be analogous to trying to understand what a human is starting from a paraplegic patient.

Now take the ways in which the verb "to describe" has come up in this book (before the current discussion). The first time it occurred, it was (on p. 2) in connection with the Land of Oz *described* in a book or musical. That land did not exist before the book (though it did exist before the musical—this intertextual relation is a good example of a behavior expanding), not in reality and not even in the imagination of Lyman Frank Baum (the book's author): it took shape in the course of writing and illustrating the book (illustration was provided by William Wallace Denslow), it was fictional in the literal sense of made up (*facio* is Latin for "to make," "to do"), and then it continued to evolve as more books followed the first one, as readers read them, and musicians wrote songs inspired by them, and singers sang those songs. (Judy Garland has done as much as anyone to expand the subject Land of Oz.)

The second occurrence (on p. 14) is where its gets tricky, since there I mention "brilliant instances of plant intelligence also described by Mancuso," and then the idea of a language-independent reality stands out convincingly (old habits are very effective at inducing convictions): the instances of plant intelligence (one is convinced) were there long before Mancuso was even born; he noticed them, and reported on them so that others could notice them too. That is how science, and specifically biology, works. But it isn't. Mancuso has chosen, no less creatively than would a great painter (not a dauber) producing a person's "live portrait," to see certain behaviors of plants as displaying intelligence, which means: he has chosen to assimilate them to other, for most of us obvious, cases of intelligent behavior. They were not intelligent before he, or someone else like him, characterized them as such; that we might be won over by his plea, and come to inhabit a world where plants are intelligent, is no indication that this world independently existed aside from the plea, and from us accepting it. To better appreciate the point by an exercise in comparing and contrasting, consider the third and fourth occurrences of "to describe" (on p. 16). Both of them apply to Aristotle: one to his description of models of intelligence—the one he officially subscribes to and the one he surreptitiously intrudes—, the other to his describing how nature makes an exception to its usually advantageous arrangements in order to allow for the scandal of the octopus. This is biology too, and we agreed to live in Aristotle's biological world for two thousand years; having rejected it now, we can see better how much of an invention it was, how little it used language to report on facts that were just there, language or no language, how the cephalopoda it *described* were as made up as the Greek gods—as self-servingly made up. (The fifth occurrence, on p. 23, is not my word but Turing's, and works like the last two: he says how the tradition "describes" rules as he is in the process of countering that description.)

Galileo was called by Einstein the father of modern science. The reason given for this assessment is interesting; let us look at it.

> [A]ll knowledge of reality starts with experience and ends in it. Propositions arrived at by purely logical means are completely empty as regards reality. Because Galileo saw this, *and particularly because he drummed it into the scientific world*, he is the father of modern physics—indeed, of modern science altogether.[16]

Stephen Hawking agrees. "Galileo," he says, "perhaps more than any other single person, was responsible for the birth of modern science."[17] And here, too, the reason given is not just that he based his research on observation and experience, but also that he wrote his masterpiece, *Dialogue concerning the Two Chief World Systems* (where he defended the Copernican view of the universe, as Bruno had done, less effectively and with a deadlier upshot, half a century earlier), in Italian rather than academic Latin, and that the book became extremely popular outside the universities. In a more critical tone, Paul Feyerabend has claimed that "the experience on which Galileo wants to base the Copernican view is nothing but the result of his fertile imagination, [...] it has been *invented*."[18] He has a point, but one that he tends to overstate, and clarifying the exact manner he overstates it is crucial in order to reach a proper understanding of the role of descriptions—scientific and otherwise.

Here is some more Feyerabend:

> The image of 20th-century science in the minds of scientists and laymen is determined by technological miracles such as colour television, the moon shots, the infra-red oven, as well as by a somewhat vague but still quite influential rumour, or fairy-tale, concerning the manner in which these miracles are produced. According to the fairy-tale the success of science is the result of a subtle, but carefully balanced combination of inventiveness and control. Scientists

[16] Einstein (1954), 271; italics added.

[17] Hawking (1988), 179.

[18] Feyerabend (1978), 81.

> have *ideas*. And they have special *methods* for improving ideas. The theories of science have passed the test of method. They give a better account of the world than ideas which have not passed the test.[19]

If Galileo invented the experience he offers in support of his Copernican view, then that view is no less a fairy-tale than the one supporting the "scientific method" of which he was one of the earliest proponents, and no less than Feyerabend's own impassioned summons to anarchy. I agree: all these tales are phrased in a language—however specialized and mathematized it might be, to choreograph gravitas and scare off unbelievers (Charles Davenport's eugenics cesspool was highly mathematized too, as are the jargons of numerologists)—thus sharing in language's irreducibly playful nature. But that is only the beginning of *this* tale, which I do not intend to continue by commandeering Feyerabend's blanket statement: "The only principle that does not inhibit progress is: *anything goes*".[20] The randomness elicited by that statement is not to be denied: some tales are able to pick up steam while others fade away; the behaviors advertised by the former expand immeasurably while those advertised by the latter shrink and (temporarily) expire; and chaos has a heavy pull on determining which is which. Still, chaotic butterflies can be helped, and it is to the *quality* of linguistic play that I would direct attention in search of a guide.

Bring back the kid again: we know that he does not have much to learn from the toy car if its controls are scanty, or from armies in battle if all he can work (to play) with is bottle caps. Same thing when one plays in language and with tales issuing from it. "Have we a clear picture of the circumstances in which we should say of a pot that it talked?" Wittgenstein asks. And I would answer: It depends. If the fairy tale he alludes to just says "Once upon a time a pot talked. The end" then No: we don't have a picture thereby, clear or unclear. But suppose the tale goes on for hours (or dozens of pages) explaining how a pot, say, bounces rhythmically on the stove emitting sounds that make melodies that influence what all other pots and pans nearby do, so that by hearing those sounds they cook meals more or less slowly, more or less deliciously; then we would begin to have a clear*er* picture of the matter, which would improve as more touches are added—the more unforeseen and yet fitting the touches, the more satisfying the outcome, and the closer we get to answering Wittgenstein's challenge. Being Italian myself, and having worked on Galileo's science, philosophy, *and language*, I can attest to the fact that this is how he was able to drum his empirical stance into the scientific world, and to make his book extremely popular outside the universities, despite the serious mistakes that are contained in it[21]: by the literary quality of his writing. Quoting from myself again:

---

[19] Ibid., 300.

[20] Ibid., 10.

[21] Galileo claimed that planetary motions were circular rather than elliptical, and that tides were caused not by the moon's attraction but by the irregular motion resulting from the combination of the Earth's rotation around itself and revolution around the Sun. On both counts, Johannes Kepler had stated the correct view long before Galileo's *Dialogue*. The latter's argument for his first mistaken view, incidentally, is a good example of the kind of conclusion "arrived at by purely logical means" that Einstein finds "completely empty as regards reality."

> Galileo is the father of the new science because he not only contributed to founding it but was also able to narrate it with that care for plot and that dignity of expression with which a great writer makes the strangest occurrences look reasonable.[22]

My tale runs contrary to the mainstream one, and it will be good to rehearse all its main twists and turns so far. It is easier and safer to play with linguistic substitutes of subjects and behaviors: instead of getting on a horse with a dumber friend and go on to explore the landscape of La Mancha, looking for giants and damsels in distress, we can more easily and safely play with words that stand for horses and friends and giants, and join them in the sentences of a story about exploring La Mancha and looking for distressed damsels. If this behavior expands in powerful ways, it may cross over into something that is no longer just linguistic: there may be people who are not just objects of discourse and who are going around some barren territory (not necessarily in La Mancha) looking for the realization of their dreams of love and acclaim. The story will thus have acquired referents: there will be something it describes other than itself; it will no longer be only weaving yarns in the texturing of a self-enclosed and self-referential tapestry—there will be something outside the tapestry, maybe coming out of it (as in Edgar Allan Poe's *Metzergenstein*[23]) that it talks about. And, though it is entirely up in the air which stories will have such fortune, and of many that do we might be inclined to say that they don't deserve it, many others show some merit, so we can say that their fortune is related to how well they are told, how much absorbing detail they spell out, how many surprising directions they take, how well they play (!) with their hearers/readers' expectations.

Some of the fortunate stories will be categorized as literature, or philosophy, or some other hypothetically harmless venture, and, though for their fans they might be among the most real portions of their lives, the general consensus will broadcast that they need to stay bracketed: that their referents belong to the "imagination" and that, if someone starts wandering in La Mancha on a horse looking for glorious trouble, he might be psychologically disturbed—that is how the spreading of play is often averted: by blaming *it* as pathological, so that pathology can rule undisturbed. But there may be times when dread brews portentously and people are ready and willing to try anything; then some of those stories that would otherwise only get a shrugging of shoulders and a complacent smile are suddenly unbracketed and their referents become real—fantasies puzzled out in a lab are likely to be given a chance. Turing's metaphorical computing machine metamorphoses into a monstrous contraption (and investment). Einstein's... Let's talk about Einstein for a moment.

General relativity was published in 1916. It was a beautiful tale; in order to so expand it that it would be incorporated into a physical description of reality it needed (Einstein himself would say) corroboration by observation and experience. The theory predicted that the light coming from stars would be deflected (more than Newton predicted) as it was passing by the Sun; and such deflection could be observed well during an eclipse. In 1919 two expeditions were conducted to make observations during an eclipse, to the Brazilian town of Sobral and to the West-African

[22] My (2008), 97; translation mine.

[23] Poe (1983), 78–86.

island of Principe. Arthur Eddington, director of the Cambridge Observatory, was part of the second one. The first one was able to take some promising pictures; in Principe observation was plagued by clouds and rain. But Eddington's enthusiasm made the difference: back in England, he made a presentation on November 6 claiming confirmation. That made the British and later the American press: on November 10 *The New York Times* titled "Einstein Theory Triumphs" and Einstein became an international star overnight. Eddington (1923) cemented that fame. As for Einstein, at least one anecdote (which, of course, has been contested—it would not fit the dominating *narrative*) reports him saying, about the possibility that the outcome had been different: "I would have had to pity our dear God. The theory is correct all the same."[24] So much for "all knowledge of reality starts with experience and ends in it."

Then the tale takes a murderous twist. On August 2, 1939, Einstein addressed a letter to U. S. President Franklin Delano Roosevelt[25] which read in part:

> Some recent work by E. Fermi and L. Szilard, which has been communicated to me in manuscript, leads me to expect that the element uranium may be turned into a new and important source of energy in the immediate future. Certain aspects of the situation which has arisen seem to call for watchfulness and, if necessary, quick action on the part of the Administration. [...] This new phenomenon would also lead to the construction of bombs, and it is conceivable—though much less certain—that extremely powerful bombs of a new type may thus be constructed. [...] In view of this situation you may think it desirable to have some permanent contact maintained between the Administration and the group of physicists working on chain reactions in America. [...] I understand that Germany has actually stopped the sale of uranium from the Czechoslovakian mines which she has taken over. [...] [At] the Kaiser-Wilhelm-Institut in Berlin [...] some of the American work on uranium is now being repeated.

This spark ignited the Manhattan Project, which starting in 1942 employed around 130,000 people and spent some thirty billion dollars of today's money (talk about monstrous investments) to build the first atomic bombs, two of which (morbid gallows humor nicknamed them Fat Man and Little Boy) were deployed (three months after Nazi Germany capitulated, and all worries about its work on uranium had vanished) on the relatively intact Japanese cities of Hiroshima and Nagasaki, to better register their effects. The virus had gotten out of the lab; testing was done in the open air; and an era of nuclear scare and blackmail was inaugurated.

Back to Einstein. His famous 1905 formula (perhaps the most famous formula in the history of science) states that energy equals mass multiplied by the square of the speed of light: $E = mc^2$. "Mass" and "energy" are then two names of the same sort of thing, which implies that one can be converted into the other *and* it implies, given the enormous conversion factor c^2 (the speed of light is over 186,000 miles per second, and its square is over thirty-four and a half billion), that an inconceivable amount of energy can be released by converting a tiny particle of matter into it. To be sure, Einstein's formula doesn't begin to give a clue on how to perform this conversion, or

---

[24] See Rosenthal-Schneider (1980), 74.

[25] See it at https://www.osti.gov/opennet/manhattan-project-history/Events/1939-1942/einstein_letter.htm.

even prove it feasible; but it opens the way to thinking of it, and eventually carrying it out. Empirically, his politics (the writing of the letter above, by the most illustrious scientist, to the most powerful person on Earth) was more influential in launching the nuclear age than any of his scientific work; but, in the transcendental scenery where we move, the connection of his disinterested, intellectual, *and brilliant* theorizing with Japanese massacres and universal anxiety is poignant. This link is indicative of what is at stake in play, and how treacherously—even in the substitute kind. There are two main aspects to the lesson it teaches us.

First, keeping play under wraps or limiting it to toys, to acting, or to words, is not, as I already noted, foolproof. Eventually, the poison (or nectar) cooked up there will spill over, and *play out* its potential to the hilt. There is no move or character devised by a poet (a *maker*—that is what "poet" means, in the Greek where it comes from: a creator of something that was not seen before) which has not, at some point, found its way into the prose of everyday life; sadly, there is also no horror conjurers of nightmares have fabricated that has not come to haunt our waking moments. Behaviors expand outside the protective walls where one tries to insulate them; what is said, or written, will inevitably once be done. Even more, *the consequences* of what is said or written, and may have been unintended or unseen by whoever did the original saying or writing, will be brought to the fore. Einstein's formula is a helpful reminder of that; nothing in it hints at split atoms or nuclei, let alone at apocalyptic explosive devices; but, once it was out there, someone would think of them, and make them happen.

When I first ventilated my warnings, ages ago, I was aware that they sounded like science fiction, and said so. "The kind of picture that emerges [...] [from this scenario] reminds one of *Invasion of the Body Snatchers*," I quipped[26]; and it was taken by some as a *reductio* of my argument. It was way too early: "serious" philosophers could still let science fiction do their work, while they blathered weary Cartesian twaddle. Now the fiction has (indeed!) invaded reality, as it was to; the consequences of what was then to be seen (but was not) have become apparent; and they don't know where to run for cover.

The second thing we learn from Einstein's lesson is that size matters. Some nasty or not-so-nasty bookish probes can find their way to actual comfortable nooks and prosper there, as real-life serial killers bury bodies in the woods in honor of the bad guys of some serial flicks or real-life retirees revamp their squalid neighborhoods by planting the exotic flowers seen in glossy magazines; none of that is going to have more than local impact and sooner or later it will be flushed out and consigned to oblivion—returning to the pages or film where it first appeared. But, when something acquires gigantic proportions, there is no avoiding the catastrophe it forecasts; and such is what we are facing with artificial intelligence—something of the same scale as nuclear energy, only worse. Not because nuclear energy can't blow up the whole world, but because that's the *only* thing it can do: destroy, or poison, all of humankind; get rid of it. Artificial intelligence, on the other hand, might not destroy humankind; it might enslave it, after sucking out *its* intelligence and turning it into a collection of

[26] My (1991), 526.

brutes. I believe that is worse. There is less dignity, less self-respect, to the destiny this prospect intimates.

We need to wait a little longer before we get to the details of how that destiny is configured; additional elements of the conceptual framework need to be worked out. But I cannot deny that the discussion of language avers with graphical might the ambivalence of play: of how innocent and lurid it can be at the same time. And some will expect me to supply a recipe for how to enjoy the innocence while skirting the filth. I must tell them that I have no such, and that I am not even assuming here the posture of one who does: I am telling a story about what is going on, not emanating statutes for changing it. I have my own, personal opinion about what *ought to* be going on; but it will be seen when the time comes that I urge no moral distinction for it—it is just what I would prefer. As for recipes, I have none—not now, at least. Back when I started on this, I would have had a modest (no foolproof!) piece of advice to venture, and I am happy to mention it here because it relates to an important topic I will discuss two chapters down. When dangers impend, it is vital to become conscious of them, to pay attention to them; and to do so as early as possible, since the only measures that might have any chance of success are those adopted early, and to move early we must be quick at realizing what is coming if nothing is done. Which is hard, as Niccolò Machiavelli tells us:

> Remember what the doctors tell us about tuberculosis: in its early stages it's easy to cure and hard to diagnose, but if you don't spot it and treat it, as time goes by it gets easy to diagnose and hard to cure. So it is with affairs of state. See trouble in advance (but you have to be shrewd) and you can clear it up quickly. Miss it, and by the time it's big enough for everyone to see it will be too late to do anything about it.[27]

His beloved Romans were good at this:

> [S]ince they had this capacity for seeing a threat in advance, the Romans always knew how to respond. They never put off a war when they saw trouble coming; they knew it couldn't be avoided in the long run and that the odds would simply shift in favour of their enemies. [...] They never took the line our pundits are constantly giving us today—relax, time is on your side—but rather they put their faith in their own foresight and spirit. Time hurries everything on and can just as easily make things worse as better.[28]

I am afraid the pundits had their way in the case we are addressing here; they aired reassuring mantras and confounded everyone; they gabbled about Cartesian consciousness (ultimately, about the holy breath) hindering what good use consciousness could have (this will become clear at the opportune place). Time has hurried on and nothing much can help any more. The only thing I will say in the pundits' defense is that this case was harder than the Romans': the latter only had to anticipate their enemies; here one had to revolutionize one's entire metaphysical tradition. To the remaining stages of this revolution I need to turn now.

---

[27] Machiavelli (2014), 11.

[28] Ibid., 11–12.

References

Aristotle. *The Complete Works of Aristotle*, edited by Jonathan Barnes. Princeton NJ: Princeton University Press, 1984.

Bakhtin, Mikhail. *Rabelais and His World*, translated by Helene Isvolsky. Bloomington IN: Indiana University Press, 1984.

Bencivenga, Ermanno. "A New Paradigm of Meaning." *Synthese* 72 (1987), 599–621.

Bencivenga, Ermanno. "The Electronic Self." In *Advances in Scientific Philosophy*, edited by Gerhard Schurz and Georg J. W. Dorn. Amsterdam: Rodopi, 1991, 517–528.

Bencivenga, Ermanno. *Il pensiero come stile. Protagonisti della filosofia italiana.* Milano: Bruno Mondadori, 2008.

Bencivenga, Ermanno. *The Logic of Mysticism*. Berlin: Springer, 2025.

de Saussure, Ferdinand. *Course in General Linguistics*, translated by Wade Baskin. New York: Philosophical Library, 1959.

Descartes, René. *The Philosophical Writings of Descartes*, in three volumes, translated by John Cottingham, Robert Stoothoff, Dugald Murdoch, and Anthony Kenny. Cambridge UK: Cambridge University Press, 1985/1991.

Diderot, Denis. *The Paradox of the Actor*, translated by Walter Herries Pollock. London: Chatto & Windus, 1883.

Eddington, Arthur. *The Mathematical Theory of Relativity*. Cambridge UK: Cambridge University Press, 1923.

Einstein, Albert. *Ideas and Opinions*. New York: Crown Publishers, 1954.

Feyerabend, Paul. *Against Method*. London: Verso, 1978.

Fodor, Jerry Alan. *The Language of Thought*. New York: Thomas Y. Crowell, 1975.

Hawking, Stephen. *A Brief History of Time: From the Big Bang to Black Holes*. New York: Bantam, 1988.

Kant, Immanuel. *Critique of Pure Reason*, translated by Paul Guyer and Allen William Wood. Cambridge UK: Cambridge University Press, 1998.

Machiavelli, Niccolò. *The Prince*, translated by Tim Parks. New York: Penguin Random House, 2014.

Montessori, Maria. *The Montessori Method*, translated by Anne E. George. Mineola NY: Dover, 2002.

Plato. *The Republic*, translated by Allan Bloom, second edition. New York: Basic Books, 1991.

Poe, Edgar Allan. *The Unabridged Edgar Allan Poe*, edited by Tam Mossman. Philadelphia: Running Press, 1983.

Rosenthal-Schneider, Ilse. *Reality and Scientific Truth: Discussions with Einstein, von Laue, and Planck*. Detroit MI: Wayne State University Press, 1980.

Saenger, Paul. *Space between Words: The Origins of Silent Reading*. Redwood City CA: Stanford University Press, 1997.

Wittgenstein, Ludwig. *Philosophical Investigations*, translated by Gertrude Elizabeth Margaret Anscombe. New York: Macmillan, 1953.

Wittgenstein, Ludwig. *Tractatus Logico-Philosophicus*, translated by David Francis Pears and Brian Francis McGuinness. London: Routledge and Kegan Paul, 1961.

Chapter 5
Thought

Abstract The traditional model of how language works has it communicate thoughts from one mind to another. There is no assurance, indeed it can hardly be believed, that the thoughts thus transmitted are the same in both minds. Frege, the mythical father of analytic philosophy, made the cavalier assumption that thoughts exist on their own, in a realm distinct from both that of ideas and that of outer things, and people (minds) just grasp them, without explaining what such grasping amounts to. For being a father of an illustrious philosophical movement, he certainly moved awkwardly around issues, including language, which was at the same time his main tool and something he declared to be constantly struggling against—without learning any lessons from the persistence of this struggle. In my view, the direction to take is not from thought to language but the other way. Thought belongs to the constitution of a private sphere: a sphere, that is, to which the public is deprived, denied, access, as an additional safety measure against too early and too risky dissemination of playful content. But private thoughts are friable: just like dreams, they are quickly forgotten. Ways must be found to fix them, for example on paper; and, as soon as that happens, they become potentially public again, and the public can encroach on them. The whole universe of concealing and discovering, of ciphering and deciphering, finds its rationale, and its huge philosophical significance, here.

Keywords Thought · Frege · Private · Hiding · Deciphering

There are two ways of looking at the Aristotle-de Saussure model of how language works. On the one hand, we see it assign to language a purely instrumental function: make it just carry something (a meaning, a concept, an image, a mental representation) from one speaker to another. I took care of that in the last chapter: language is not subservient to anything else; it is an autonomous space of play, where expansions of behaviors can be tried and tested with greater safety than in more robust media, and then either gradually and cautiously spread to those other media or unwittingly let go, possibly to make everything crumble. There remains the other angle to take up: what language is supposed (by Aristotle, de Saussure, and their lackeys) to be subservient to. In the end, I want to say that the relation of subservience is (as many

E. Bencivenga, *The Philosophy of (Artificial) Intelligence*,
https://doi.org/10.1007/978-3-032-09347-9_5

traditional relations here) to be reversed; but, before I get to the end, let us inquire into how the privileged party of this master-slave relation has been understood so far.

The mythical father of analytic philosophy is Gottlob Frege; "mythical" because this parenthood has no basis in history. Frege, a mathematician, worked in an obscure corner of German academic life, was snubbed when he tried to connect with his far more redoubtable colleague David Hilbert, and his only substantial contact with analytic philosophy, by way of Bertrand Russell, one of its founders, was an unmitigated disaster: Russell alerted him to an irreparable flaw hiding in plain sight at the foundations of his system and he never recovered from the shock—he published nothing for fifteen years and then only produced three short essays not long before his death. But Russell, despite handing this fatal blow, was a great admirer of Frege, as was Wittgenstein (which counted even more, since Russell was an even greater admirer *of him*), so eventually Oxbridge came around and began an appropriation of the old fart. In 1952 Blackwell published *Translations from the Philosophical Writings of Gottlob Frege*, by Peter Geach and Max Black;[1] one of the later essays, which had not made it into the Blackwell collection, appeared in 1956, translated by Anthony and Marcelle Quinton, as "The Thought" (German "*Der Gedanke*") in *Mind*, the very cradle of analytic philosophy (Russell's friend George Edward Moore, another recognized founder, edited it for twenty-six years). Mythical projection was thus accomplished and paternity established, so extravagant praise could follow. In an article published in *The New York Times* on August 14, 1977, by Taylor Branch, the Columbia University sage Sidney Morgenbesser, better known for his witticisms than for any bright speculations, is quoted as ranking his colleagues, alive or dead, "according to four categories: Gods, Geniuses, Bright Guys and a catchall group called Competent Philosophers." Of Gods he only allowed Plato, Aristotle, and Kant, but a footnote adds: "Morgenbesser later declared that he had, upon reflection, added a fifth category called Minor Deities and that he conferred that distinction upon Gottlob Frege" (no one else is mentioned as belonging there—not Descartes or Hume or Hegel or Nietzsche). Moving from sophomoric humor to daily fodder, the Wikipedia article on Frege chronicles that he "is widely considered to be the greatest logician since Aristotle [I would pick Gödel for that honor, or Turing, since their stuff at least did not self-destruct] and one of the most profound philosophers of mathematics ever," as well as, of course, "the father of analytic philosophy."[2] It

[1] Frege (1952). But the first translation of works by Frege was into Italian: Frege (1948).

[2] The hype extends to clumsily attempting to protect mythical daddy from misunderstanding, by sleight of hand if needed. The following is a telling example. Frege used the perfectly ordinary German word "*Bedeutung*" and its cognates, which mean "meaning" and its cognates, to express such peculiar views as that the meaning of the numeral "2" is the number 2 and the meaning of a sentence is its truth-value (no more peculiar, I should point out, than his other view that a sentence is a singular term, of a kind with proper names and definite descriptions). In their 1952 translation, Geach and Black, evidently at a loss, translated "*Bedeutung*" as "reference," thus obscuring the fact that saying "The meaning of a sentence is a truth-value" is making a strong and questionable philosophical statement, on which one might or might not agree, rather than exploiting a term of art on which no one else could have a say. In their 1979 translation of the *Posthumous Writings*, Peter Long and Roger M. White opted for the translation "meaning," choosing exegetical neutrality and letting the reader make up her own mind (in English, as much as she would have to in

sounds like the kind of authority I could reliably take my lead from; I will do so based on the piece translated in *Mind*.

There are, Frege says there, three realms. One contains things in the outer world, which can be perceived by the senses but are independent of being perceived: they exist on their own, whether or not anyone sees them, smells them, or touches them. Then there are ideas (the German word is "*Vorstellung*," the same one translated in Kant as "representation"; I believe that "idea" is a bad choice): each of them has a single bearer it depends on (no bearer, no idea); the idea "belongs to the content of his consciousness" (299); and "no two men have the same idea" (300; the German word for that "men" is "Menschen," which does not mean "men" but "humans"—there goes the heterosexual pair of translators).

> I notice a strawberry among the green strawberry leaves. My companion does not notice it, he is colour-blind. [...] Now does my companion see the green leaf as red, or does he see the red berry as green, or does he see both as of one colour with which I am not acquainted at all? These are unanswerable, indeed really nonsensical, questions. [...] For it is impossible to compare my sense-impression with that of someone else. (299-300)

This is where transcendental realism floundered in its project of building an epistemology: a doctrine of knowledge (thus opening the way for Kant's Copernican revolution). For how can we account for the possibility of us acquiring justified true beliefs about independent things in the world[3] when all we have to go on is private experiences that cannot be compared with anyone else's, even less shared with anyone else, and certainly cannot be compared with independent things out there? But have no fear, for Frege is coming to the rescue! He begins by astutely identifying the problem:

German—though I remember hearing talk at a German conference of Frege's *Referenz*, a backward translation from a bad English translation that might have assuaged the German spirit, too). But here comes Bell (1980), defending the choice of "reference" with arguments like this: "what could be more unattractive or confusing to the newcomer than to discover that a philosopher revered for his incisive clarity of style and for his profound insight into how language works, seems to believe that the sentence 'no men are mortal' has the same meaning as the sentence '2+2 = 5'; or that every sentence of fiction is quite without meaning; or that the predicate 'is a round square' is perfectly meaningful, while the predicate 'is a christian' is in fact meaningless? It was presumably to avoid just such consequences as these that every translator, up to the present time, has avoided rendering '*Bedeutung*' as 'meaning'" (195). Therefore, when necessary, incisive clarity of style and profound insight will be manufactured at will. As a sad footnote to this discomfiting episode, Long and White (1980) point out, in their response to Bell, that "the Geach and Black translation of some of Frege's writings has been revised in accord with our translation of '*bedeuten*' and its cognates" (201). Which was true, indeed, of the third edition (1980) of that storied translation, but this decision, rather than bringing about a universal agreement on the obvious, opened the gates to total confusion, as some translators did adopt "meaning", some stuck with "reference," some preferred "denotation," and some even decided to leave "*Bedeutung*" untranslated. Nice way of doing justice to the old man's incisive clarity of style—real or fictional as it might have been!

[3] This definition of knowledge goes back to Plato (1992), 80 (201d). It was challenged by Gettier (1963)—and, arguably, by others before him—giving rise to a whole industry of efforts inconclusively aiming at finding a "fourth condition" or some other way out of the swamp. I will do nothing with this whole line of wasted work other than register its failure.

> If every thought requires a bearer, to the contents of whose consciousness it belongs, then it would be a thought of this bearer only and there would be no science common to many, on which many could work. But I, perhaps, have my science, namely, a whole of thought whose bearer I am and another person has his. (301)

And then, to save the day, he *postulates*, arbitrarily, that a "third realm [*ein dritter Reich*[4]] must be recognized" (302):

> What belongs to this corresponds with ideas, in that it cannot be perceived by the senses, but with things, in that it needs no bearer to the contents of whose consciousness to belong. Thus the thought, for example, which we express in the Pythagorean theorem is timelessly true, true independently of whether anyone takes it to be true. It needs no bearer. It is not true for the first time when it is discovered, but is like a planet which, already before anyone has seen it, has been in interaction with other planets. (ibid.)

No ground can be offered for the Aristotelian claim that the affections of the soul "are the same for all" (Frege has seen that, and has rejected the claim), nor for conferring Saussurian concepts or sound-images any but the lamest, most unhelpful agreement between two minds; and, with such a level of uncertainty spoiling the broth, can we still presume that those two have provided a model for how something (one and the same thing) is made common, shared, by two speakers? Frege faces the difficulty by dividing it. First, he just asserts (in passages like the last one above) that there can indeed be a "science common to many, on which many could work" because there are identical entities (his "thoughts"), which are neither outer things nor inner ideas but take some features from both, and it is these entities that speakers and scientists share, that are common to them. It's a hard bite to swallow, but in metaphysics we are used to similar fare: Fregean thoughts are no more fanciful than, say, Platonic ideas or Leibnizian monads or Schopenhauerian will. Problem is: despite this epitome of metaphysical profligacy, the Fregean division does not conquer, since we still have to explain *how* different speakers, or scientists, get to think, or talk about, the same thoughts, and work on the same science. (The Kantian question "How is it possible?"—the only one making philosophy go forward—has not been addressed yet.) What does Frege do to that? He brings in a *word*:

> Not everything is an idea. Thus I can also recognize the thought, which other people can grasp just as much as I, as being independent of me. I can recognize a science in which many people can be engaged in research. We are not bearers of thoughts as we are bearers of our ideas. We do not have a thought as we have, say, a sense-impression, but we also do not see a thought as we see, say, a star. So it is advisable to choose a special expression [!] and the word "apprehend" offers itself for the purpose. A particular mental capacity, the power

[4] Though Frege died before Hitler's rise to power, this expression is involuntarily revealing, as he was, at least at the end of this life, a rabid racist and antisemite. For regret expressed by a devout follower on this heinous attitude, see Dummett (1973), xii, which also informs us that the fragment of a Frege diary where it is clearly stated was expunged from Frege (1979)—the same volume that started the fuss noted above on the translation of "*Bedeutung*." The original German edition which Frege (1979) translates was published in two volumes, in 1969 and 1976. Frege (1979) only translates the first volume; the second one was partially translated as Frege (1980); and neither volume contained the diary, which only appeared twenty years later in Mendelsohn (1996). All efforts should be made not to obstruct the beatification process!

of thought, must correspond to the apprehension of thought. In thinking we do not produce thoughts but we apprehend them. (307)

The image of a word that offers itself (a literal translation of the German "*bietet sich*"), as a sacrificial lamb, is endearing, and is also glum, as the word is slaughtered to no avail: after digesting it, we are not an inch closer to getting the notion of what Frege intended to be going on here—or to admitting that he intended anything at all. The Quintons do a canny job of selecting a Latinate translation for the German verb "*fassen*" after the sacrifice is performed: in "to apprehend," and in the cognate substantive "apprehension" (for which Frege uses a substantivized form of the verb), an English speaker will rarely hear the Latin "prehendere" (to grasp, to take) it comes from—both verb and noun have by now been sequestered to mental contexts and reserved for mental operations. The Quintons slide over the fact that, before the word generously "offered itself," it had already occurred, in the same passage, and had been given by them its ordinary translation "to grasp" ("other people can grasp just as much as I"). Later translations have avoided this walking on eggshells and have chosen "to grasp" all the way; but 1956 was an early, still fragile phase of mythmaking, and one was well-advised to go easy on it. As for Frege, this is what he has to say on the matter, in a footnote:

The expression "apprehend" is as metaphorical as "content of consciousness." The nature of language does not permit anything else. What I hold in my hand can certainly be regarded as the content of my hand but is all the same the content of my hand in quite a different way from the bones and muscles of which it is made and their tensions, and is much more extraneous to it than they are. (307n)

One first comment I make on this passage—brace yourself: it's going to last a while—has to do with the complaint it expresses about language, curious in an author who has been lauded (among other, already cited, ecstatic tributes) as a monumental figure in, or even the initiator of, analytic philosophy of language. Similar complaints occur everywhere in his work—though many of them, as we will see, are to be found in unfinished and unpublished notes written during his long period of silence (and religiously edited and translated by recent worshippers). Language, he alleges, is misleading:

[M]ost logicians, misled by language, take the words "some numbers" in the sentence "some numbers are prime numbers" together, and treat their meaning as a logical subject of which the property of being a prime number is predicated, just as it is predicated of say the number 2 in the sentence "two is a prime number."[5]

So language brands a concept as an object, since the only way it can fit the designation for a concept into its grammatical structure is as a proper name. But in so doing, strictly speaking it falsifies matters.[6]

There is, however, no hope of fixing this inadequacy:

---

[5] *Logik*, in Frege (1979), 239.

[6] "Über Schoenflies: *Die logischen Paradoxien der Mengenlehre*," in Frege (1979), 177.

> [O]n thorough investigation it will be found that the obstacle is essential, and founded on the nature of our language; that we cannot avoid a certain inappropriateness of linguistic expression; and that there is nothing for it but to realize this and always take it into account.[7]

Which is aggravating, because we have no other tool than language for the sort of work Frege is doing (logic, philosophy of math, philosophy of language, whatever):

> These investigations are especially difficult because in the very act of conducting them we are easily misled by language; by language which is, after all, an indispensable tool for carrying them out.[8]

And, in carrying out this difficult task, the thesis that we understand each other appears now not as one we can prove, or even assuredly state, but only as one we must presuppose in order for the whole undertaking to make any marginal sense:

> [W]e cannot come to an understanding with one another apart from language, and so in the end we must always rely on other people's understanding words, inflexions, and sentence-construction in essentially the same way as ourselves.[9]

More than a useful resource, language is an impediment constantly in the way, blocking view of what we are really after:

> [A] battle with language takes place and I am compelled to occupy myself with language although it is not my proper concern here.[10]
>
> [I]nstead of following grammar blindly, [...] [the logician] ought rather to see his task as that of freeing us from the fetters of language.[11]

Where this liberation is ideally—but only ideally—projected into an imaginary, at long last propitious future:

> If our language were logically more perfect, we would perhaps have no further need of logic, or we might read it off from the language. But we are far from being in such a position. Work in logic just is, to a large extent, a struggle with the logical defects of language, and yet language remains for us an indispensable tool. Only after our logical work has been completed shall we possess a more perfect instrument.[12]

What is curious is not just that a "philosopher of language," whether monumental or not, should find that language is not his "proper concern" and that he judge it rather to be a nuisance one must keep at bay or a trap one must avoid falling into, while he dreams of streamlining it into something that it is not. (The regimentation of language has indeed been a major project of analytic philosophy, beginning with that "paradigm of philosophy" which was Russell's theory of definite descriptions.[13]) It

[7] "On Concept and Object," in Frege (1952), 55.

[8] "Zahl," in Frege (1979), 266.

[9] "On Concept and Object," in Frege (1952), 45.

[10] Frege (1956), 298n.

[11] "Logik," in Frege (1979), 143.

[12] "Meine grundgelegen logischen Einsichten," in Frege (1979), 252. These and other Frege passages were also discussed, for different but related purposes, in my (1993), 21-23.

[13] For Russell's theory, see his (1905). For the quoted judgment on it, by Ramsey, see his (1931), 263n.

is also that he should not reflect on this predicament, and draw no teachings from it: that it should not bring him to see language in a thoroughly new way. But perhaps for doing that we would have needed someone who was in fact a philosopher, even a midsize, purely competent one—not a minor deity in a backdoor, discount pantheon.

I said in Chapter 3 that I prefer to hassle with a language that resists me,[14] and that later it would become clear why. What "later" was hinted at there is the discussion in Chapter 4, after which we can say that, if language is our main site for a play that violates established norms and habits, then language and struggle are intimate: language struggles with any behavior that might be entrenched, and we, insofar as there is an entrenched fabric to our behavior and we are keen on remaining faithful to it, struggle with language. No wonder that, long before the crazy analytic-philosophy regimenters set up their Procrustean beds, there would have been so many and so stringent rules enforced on language: of grammar, of syntax, of word choice—more than on any other human activity. And no wonder that language has unceasingly infringed on those rules: that it takes words for a ride (the word "apprehension," for example), that it makes us forget about the subjunctive (an editor of a major academic press did not recognize it in one of my books), that it mutates spelling and contravenes etiquette. Its tumultuous romp is life itself; it is soul, if only in the vicarious domain of theatrical performance; and it is always on the verge of doing a Grotowski on the audience, of polluting the latter's stable routines, of forcing them away from themselves—having them step into yet unheard-of and unknown (and dangerous) phases. Frege's whining about it belongs to a taskmaster who would want to see everyone march in step, sing in unison, unremittingly obey orders, and can't quite figure out why there is so much immoderacy and indulgence among the troops.

Which calls for an aside on my language: the one I have been airing here (and elsewhere). Moronic analytic philosophy does not just hallucinate a nightmare where people say to each other (or think, maybe in mentalese) "The cat is on the mat" or "Whales are mammals"; it also does its best to act it out. Its agents speak and write in a corporate lingo, full of acronyms (not Language of Thought but LOT, and LOTH if you add Hypothesis to it—who knows what they are going to do, what marvelous fun they are going to have, during the time they gain by saving those few syllables?), useless formulas, and copious bibliographical references (to each other, to maximize "impact"), thus making sure that they find a comfortable place within a circle of insiders who only need sparse leads to immediately fall into notorious, well-rehearsed ruts. Their wistfulness (for they are alive too, if barely, and they too regret, if subliminally, the sorry condition to which they have condemned themselves) impetuously erupts with no rational justification (no justification *they* could regard as rational) when one of them does a con job of recycling old hat in colorful images and alluring verbiage (making, say, something as dry as an algebra into a collection of possible worlds and relations of accessibility among them); then

[14] That statement referred to the same inconvenience Frege denounced above, of having nouns (he uses "*Eigenname*," translated as "proper name," to cover a large semantical field) stand for something that is not an object. But our attitudes toward this "inconvenience" are quite divergent.

they, too, must acknowledge the emancipating eminence of words and, free for once from their fetters (not the fetters of language: the ones they were trying to impose on language, and on themselves), show their gratitude to the cheating liberator by proclaiming his genius.[15]

I don't speak, or write, their language. If theirs performs stolidity and torpor, mine performs mischief. When I sit down to write (for instance, to write *this*), I have a general idea of what I am about to say, but then words take over, they hear and repeat resonances, they multiply like grains on a sandy beach, they draw their own picture by mixing their hues. If I am right about what it is to describe (*that is*, if you buy my story about it), then a new scientific or philosophical theory is not describing (better than the competition) something that was already there, waiting to be (more) correctly described: rather, it brings something forth that, if not for its talking about it, would never have been. The "reality" it invokes was no more real before this sorcery than the Land of Oz was before Baum started writing about it and Denslow started illustrating it—or than a rugged, uneven Moon was before Galileo's *Dialogue*. So it is important that there be lots of words, not necessarily novel but combined in novel ways, used in unexpected contexts, arranged in elegant architectures, in order to make the reader's imagination (her faculty for making images) echo with mine—after all, even with someone like Kripke, it is verbal facility and luxuriancy that attempts to exonerate his obvious borrowings. (For a truly towering example, think of Plato: officially a bitter adversary of rhetoric while he was a supreme master of it, and could enchant us by reciting fantastic narrations about prisoners in a cave, a staircase to heavenly beauty, or souls growing wings and taking up flight.) If that operation works, then there will be, for a while, a reality to be described by the new tale: one where Venus is no less horned than the Moon, or where the mass of the Sun deflects the light coming from distant stars.

The essence of my second comment on the Frege quote of p. 59 is that he doesn't do any of the above. There is nothing wrong with bringing in a word (though there might have to be for him, given how much he disparages words and language), since words are the building blocks anyone utilizes here, and there is nothing wrong with that word being a metaphor, with us being asked to grasp something that cannot be touched: the *word* "metaphor," from the Greek "*metaphéro*," means "transference"; and, when we start building our "description" of what is not yet, there is no alternative to forcing new wine into old bottles—new meanings into old words.[16] But he doesn't go on from there to building anything, and the metaphor, from hopeful which it could have been, gets stale. For we would want to know: how does someone who grasps a thought differ from one who doesn't? how would we determine that someone has *not* grasped a thought? how could we argue that two people have grasped the same thought? No answers to these questions are coming, so the story of grasping thoughts

[15] In the *New York Times* article quoted on p. 56, Saul Kripke (the person evoked in my text) is said to be "the one genius of our [philosophical] profession" (Robert Nozick speaking). I have called in question this label in my (2024a, b).

[16] This, I argued in my (1987), is what Kant does when carrying out his revolution, and a main reason why he is so hard to read.

looks much like the one that only said a pot talked, and nothing else. It doesn't give us "a clear picture of the circumstances" in which a thought is grasped; no description of a "reality" matures from it.

The comparing and contrasting got a bit wayward, though in the process we learned some additional useful detail on the incorrigible effrontery of language; let me hurry now to tell my story (to be compared and contrasted with the mess above). I begin with a word that was already used a few times: "private." In the Merriam-Webster dictionary, the first class of definitions for it stress, in a positive vein, who or what is supposed to have access to a private X ("private" is predominantly an adjective; the only use of it as a noun is to refer to an army private): "intended for or restricted to the use of a particular person, group, or class"; "belonging to or concerning an individual person, company, or interest"; "restricted to the individual or arising independently of others." The second and the third classes bring out, negatively, who or what is supposed to be excluded access to a private X: "not related to one's official position"; "not holding public office or employment"; "not known or intended to be known publicly" (here is a reference to "secret"). The first class, and especially "belonging to or concerning an individual person, company, or interest," captures the sense of "private" that is current in Cartesian-like projects: my ideas (as well as my volitions, imaginings, and feelings) belong to the individual me, and it is an open question, to be settled by one of those projects, whether anything that is *not* private (a chair that anyone, not just me, can sit on; a warmth that anyone, not just me, can feel) corresponds to them—in an unpacking of "correspond" that is likely going to be different for different such projects. I will eventually get to something like that sense, but I *start* at the opposite end, emphasizing the negativity of the second and third classes, especially of "not known or intended to be known publicly," and connecting it to the etimology of "private," which comes from the past participle "*privatus*" of the Latin verb "*privo*," meaning "to deprive," "to rob," "to free." Private, for me is, first and foremost, something deprived.

Deprived of what? And by whom? Say that a government wants to spy on its citizens, and for this purpose establishes an agency that is to collect data, to organize them, and to identify people who are potential sources of trouble. It would be counterproductive to the agency's operation if it were known publicly: then the citizens, more so if they were potential sources of trouble, would avoid giving the agency any information about themselves, maybe even get their friends and acquaintances to do the same. For its purpose to be attained, then, the government will *deprive* the citizenry of access to the agency's real function, its workings, and its results. This entire business will be segregated to a *private* area, as far as the citizenry is concerned. Following upon the reference made above to "secret," it will be secret to the citizens—a case of that secret police which is run by virtually every government.

This is the model of privacy I apply to individuals. Say that a human behavior is undergoing a powerful expansion, mostly linguistic but not only so. Some outrageous stories are being told, some of the moves "described" by them are tried out. But it's too early for that ruckus to be spread far and wide, or it's never going to advisable that it be: an artist wants to keep her number clandestine before she explodes it in front of an audience as a surprise, for maximum effect; or a vicious killer is thinking out, and trying on a pillow in his bedroom, unprecedented genres of noiselessly strangling a victim. Then the expanded behavior must be protected, temporarily or permanently, from inspection on the part of anything other than the behaving subject, *depriving* anything other than that subject of access to it, isolating it in a *private* area that is secret to anything other than that subject.

As I said, the behavior that is thus made private can in principle be of any kind. Some people will dance before a mirror when they are alone, and might lock the door so as not to be embarrassed if a stranger butted in; others will sing aloud when they drive by themselves (and would be embarrassed if someone in a passing car saw their faces oddly distorted). But, for humans, by far the most private (expanded) behaviors are linguistic ones: if language is a safety measure that permits adventurous play with reduced risk, making language private affords an *extra* safety valve—language shields a speaker from the risks of rock climbing by having her only *talk* about rocks and climbing, and making language private shields her from the risk that an unsympathetic bystander become aware of her rock-climbing plans or wishes, and make plans of his own to stunt hers. For humans, language made private is *thought*; and the reversal I forewarned about is now apparent. Whereas in the Aristotle-de Saussure model one first has something in mind (an affection, a concept, a sound-image—in the Cartesian ample sense of thinking, a thought[17]) and then uses words to transmit it to another, here the kind of sharing that language provides comes first—as part of the sharing that issues from any interacting behaviors—and then some of this language is deliberately *not* shared: fences are raised to keep it aloof, dark screens are put in place to occlude view of it. So I do get to what is private to me belonging only to the individual me (which is where the opposition starts from); but I get there as the result of *denying* entrance into that zone to anyone else—the negative sense of "private" is primary here.

Frege's amateurish shallowness has some advantages. A major one is that, making thoughts objective—indeed, timelessly so—by fiat, it buys (or, rather, steals) a solid, complex structure for them. The thought of the Pythagorean theorem sits impassive in the third realm, untouched by the goings-on in the cluttered biography of whoever grasps it, and maintains its patent links to, say, Euclid's First Theorem (from which it is proved in Euclid (1956)) or to the incommensurability of the side and the diagonal of a square (which follows from it)—whether those links are patent to anyone or not. That it be grasped haphazardly and fleetingly by anyone makes no difference to its nature: whenever someone thinks it (that is, grasps it), *what* it thinks is this very entity, with all of its ontological firmness and all of its logical relations (of which, of course, the person grasping it is not also thinking—not every one of them for sure). For those who don't want to exploit this advantage "of theft over honest toil,"[18] life is much harder, as to them a thought is, to begin with, only something that occurs

---

[17] See Descartes (1985/1991), vol. II, 19: "But what then am I? A thing that thinks. What is that? A thing that doubts, understands, affirms, denies, is willing, is unwilling, and also imagines and has sensory perceptions."

[18] The expression is Russell's, in his (1919), 71, and refers specifically to "[t]he method of 'postulating' what we want" (ibid.). To be fair, I would presume that, though "*Der Gedanke*" was published the year before, Russell would not have seen his admired Frege use that very method in it—what

in a mind, and hefty toil must be exercised (with no success so far) to grant it any objectivity: to sanction any claim that it may get over the condition of a personal whim.

I am not going to harp on the inevitable tendency of Cartesian mentalists, and transcendental realists in general, to slip into solipsism—or into what Kant labeled empirical idealism. My Kant books[19] have drawn the map of that trail of tears. What I am interested in here is the haphazard, fleeting character of the subjective experience of thinking mentioned above; what I do want to harp on is the simple, but almost universally unrecorded, fact that on this basis it is impossible to construct anything of substance.

In the bizarre account provided by conventional wisdom, I first think of something, then say it outloud or write it down, and then someone else (a hearer or reader) acquires something from it: the same something, a corresponding something, or whatever else the outcome of receiving my utterance might be according to various ministers of this bizarre gospel. Such a scheme of things could be regarded as plausible in the inane cases the high priests of the sect focus on, and feel relieved and enraptured when they have accounted for. If I were ever to think something as dopey as "The cat is on the mat" or "Whales are mammals," I would probably be able to retain it for the length of time it takes to say it outloud, to have someone hear me, and to see that person nod, or frown, or look at me in bewilderment as if my sanity was now an issue. If I were one of those priests, therefore, I would definitely feel relieved and enraptured, having thus ascertained that my scheme worked.

But now imagine a case in which a more real-life relation between thinking and speaking (or writing) is under scrutiny. I, say, am trying to prove a new theorem, or composing a new chapter of a mystery novel. I start out with an intuition of how it should go, and maybe envision one or two steps in the process; but I cannot hold on to them for more than the shortest time, nor can I do so as I move on to further steps. There is a dreamy quality to thought, and dreams are easily forgotten, though what memory we have of them, Sigmund Freud notes, is our only evidence that we had them: "we only know dreams by the memory of them after we are awake."[20] Most irritatingly, often we remember *that* we had a dream but cannot remember any of its content—just as we remember that we had the bright idea of an ingenious step in a proof or of an abrupt twist in a plot but cannot at all remember how it went. What should we do then? With dreams, Freud advises, we should write them down as soon as we wake up (do so even in the middle of the night, if we happen to wake up then, since by morning the dream will be forever lost); with bright ideas, the strategy should be the same. Don't trust that you will be able to recover something you thought once when you need it later: your mind is not a repository where goods are stored but a tempestuous kaleidoscope of ephemeral simulacra, where it's hopeless to surmise that any durable order be forthcoming. Hume put it right when he stated:

with there being a war between England and Germany, and with Russell being imprisoned for six months in 1918 for his pacifist activism (during which period of time he wrote the book cited here).

[19] My (1987, 2007).

[20] Freud (1953/1974), vol. IV, 43.

> The mind is a kind of theatre, where several perceptions successively make their appearance; pass, re-pass, glide away, and mingle in an infinite variety of postures and situations. [...] The comparison of the theatre must not mislead us. They are the successive perceptions only, that constitute the mind; nor have we the most distant notion of the place, where these scenes are represented, or of the materials, of which it is compos'd.[21]

The implication of this picture of what thinking, and the thinking mind, are is that they are frail, and that on their fractured threads nothing can be engineered, in language or in Frege's first realm, that will last more than a passing breath—"a passing shadow."[22] If you want to go the usual route, thought → language → outer things, then just think up burps like "The cat is on the mat," quickly, and as soon as you are done belch them out, and see what your audience is going to do with that valuable information. If it's something more meaty that you want to deal with, don't even attempt it: you will have lost the beginning of it by the time you are in its second paragraph.

Moving to my side of the equation, frailty continues to matter. Say that I consign a move, or more likely a speech, to privacy. I do so in order to protect it from eavesdropping or other forms of witnessing: to avoid it becoming incorporated in other subjects and behaviors. But soon, if that is all I do, the move or speech will no longer be retrievable by *my* subject and behavior; I will have lost its scent. To inhibit that debacle, I must keep a record of it: make a tangible trace of it part of my behavior, so that if need be I can pick up the trace and restore what was hidden, to others and (involuntarily) to myself. And here glitches crop up, of profound (but underrated) transcendental significance.

A tangible trace is no longer, in and by itself, private, so the only way of preventing my thought from becoming publicly accessible is to restrict access to the trace: to make *it* private, too. Except that there is no absolutely secure way of doing so, other than keeping it unsaid and undone (that is, keeping it in thought only, which would duplicate the original problem). There are a number of halfway measures; all have weaknesses; and all can fail. One can stash away the trace in a vault, to be open by a combination which, however, must also be remembered, and for which therefore one must also leave a trace. Early in Hitchcock's movie *Marnie*, the protagonist's employer Sidney Strutt keeps the combination to a safe, which he always forgets, in a drawer; soon Marnie discovers it and walks away with ten thousand dollars. Or think of how many passwords people have, how they couldn't possibly remember them all, and hence keep them in a dedicated file; all one needs to penetrate their privacy is to get hold of that file. Alternatively, one can bury the trace underground, at the risk of it being lost to himself and everyone else. In Poe's story *The Gold-Bug*,[23] a humongous treasure buried by the pirate Captain Kidd is accidentally (and

[21] Hume (1978), 253.

[22] Psalm 144: 4, in Metzger and Murphy (1991), OT796.

[23] Poe (1983), 806–836.

ingeniously) discovered; however clever and fortunate the enterprise is, one thing is clear—the treasure was not retrieved *by Captain Kidd*.[24]

A second halfway measure is also of great concern for Poe, who discusses it in both fiction (including *The Gold-Bug*) and nonfiction, and judged himself a master of it—or, rather, of disarming it. It made a brief, oblique appearance in this book before, in talking about Turing's "intelligence" (what a word!) work. It consists in enciphering what is private in a secret code, that needs a key to be deciphered or decoded; and the key, much like the combination opening the vault, is something else that must also be kept private. This measure is suggestive of a valid stretching of the notion of privacy, which can apply (not just to an individual subject, but) to a behavior including two or more subjects: a group of people that keep certain actions or texts secret to the outside and communicate with each other about them by using a code. The notion of a private letter is relevant here (in past centuries, when letters in transit could be stolen by robbers, some parts of them were enciphered, and the cipher was sent by a separate letter, in hopes that not both of them would be caught, or at least not caught by the same people), as do spy stories and, more seriously, the vicissitudes of someone like Antonio Gramsci, who did most of his writing in a Fascist prison, bound to it being read by censors, and always used pseudonyms for Marx, Lenin, or Stalin so that those names would not be recognized and crossed out.

To better appreciate what precedes, consider the extent of the reversal I am making from tradition. Early in the Second Meditation, Descartes is looking for a solid basis on which to rebuild the edifice of knowledge, after tearing it down by systematic doubt:

> Archimedes used to demand just one firm and immovable point in order to shift the entire earth; so I too can hope for great things if I manage to find just one thing, however slight, that is certain and unshakeable.[25]

And, shortly thereafter, he finds that sturdy rock in a direct contact with himself—with his own private thoughts:

> So after considering everything very thoroughly, I must finally conclude that this proposition, *I am, I exist*, is necessarily true whenever it is put forward by me or conceived in my mind.[26]

Despite the bewildering confusion about how the *cogito* should be understood, and how it is supposed to work—as an inference, a performance, or a dialectical

---

[24] Alan Turing is once again relevant here (as well as to the next halfway measure). During the war, he got worried about a German invasion and what it could do to his welfare; so he purchased two silver bars weighing together some 150 pounds and buried them near Bletchley Park (where he worked for the government as a decoder of German secret messages), enciphering their (two) locations for later recovery. But, when he went back to retrieve them, those ciphers did not help him find the buried treasure; as a result, the silver was irretrievably lost to him. For this story, too, see Hodges (1983).

[25] Descartes (1985/1991), vol. II, 16.

[26] Ibid., 17.

argument[27]—this Augustinian line[28] was to have tremendous following: as I will expound in the next chapter, a whole world (of consciousness) was to be elicited from it. In stark contrast with such a tradition, the private contact with oneself that goes under the name of thinking is for me anything but sturdy and firm: it is always about to fade, and can only recover some vigor by an anaclitic relation to something that is *not* private, and that is constantly threatening to give the game away.

There is more. Foggy and fragmentary as they might be, contents can only remain private by expending energy in keeping up defenses. The public (that is: uncontrolled spreading) is forever pressing at the door; and it is tiresome never to stop resisting that pressure. Here is another analogy with Freud: with his return of the repressed. To avoid a total breakdown of the walls, as a consequence of total exhaustion of resources, repressed material will be freed in situations where it might do little damage: dreams, jokes, artistic production. Analogously, the powerful urge to disseminate what is in the vault, or buried underground, will be given a chance when there seems to be little risk involved. The story of King Midas' barber I mentioned in the first chapter is a case in point, and like cases of unloading to a friend or to a stranger, with the alibi perhaps of being drunk, are daily variants of that ancient tale. We are all too aware of the fact that such cases often go wrong, and I won't belabor it. I will rather end this chapter by briefly limning a grandiose confusion of public and private, and consequent invasion of the latter by the former, that has been taking place under our eyes in the last couple of decades.

The qualification "social" in "social media" should be a clue: alert users to the fact that they are moving in the agora, that they are spilling their beans in society. But everything else in the experience they are having most often points in the opposite direction: they are alone with their totemic gadget; there is silence all around (or, if there isn't, and they are in a crowded subway, they can make silence by wearing earplugs); it may even be dark, as they are losing sleep in order to press keys and stay in the magical world inhabited by their "friends." None of those pointers is accurate: they are, in fact, shouting their secrets to all and sundry; but cultural evolution, as Lorenz taught us years ago,[29] goes much too slowly for technological advance, and we may well end up crushed by that chronological lag. They certainly are: as they divulge their most closeted caprice, relieved that they don't have to make an effort to hush about it, they are taking off their clothes in public and showing their private

[27] See for example Hintikka (1962) and Curley (1978).

[28] Augustine (1984) writes: "the certainty that I exist, that I know it, and that I am glad of it, is independent of any imaginary and deceptive fantasies. In respect of these truths I have no fear of the arguments of the Academics. They say, 'Suppose you are mistaken?' I reply, 'If I am mistaken, I exist.' A non-existent being cannot be mistaken; therefore I must exist, if I am mistaken" (459-460). Descartes was made aware of this connection by one of his correspondents and, as regularly happens in such cases, tried to minimize his lack of originality, and even to draw some windfall from it: "In itself it is such a simple and natural thing to infer that one exists from the fact that one is doubting that it could have occurred to any writer. But I am very glad to find myself in agreement with St. Augustine, if only to hush the little minds who have tried to find fault with the principle" (Descartes (1985/1991), vol. III, 159).

[29] See his (1974).

parts. It's not just that there are vigilant administrators listening to what they say, and using it to sell them merchandise or maybe even sell them to the authorities—that is definitely happening, and is horrific. It is also that their "friends" are none of that, that their group has no privacy, that there are all sorts of inroads into it, and that their solace of bringing down fences is paid dearly with the disappearance of their private space.[30]

Long story short: Thought is not the hallmark of a "thinking thing" that, perfectly assured of itself and its thinking, endeavors to build a bridge to what is not itself and is not thinking. It is a weak protective device to isolate potentially dangerous material and prevent it from injuring its carrier and others. Its weakness makes it forever prey to failing in its mission; its being aggrandized into the foundation of personal identity, or even into an imperial reality on a par with the whole of the "material" world, is the very essence of wishful thinking—of a thinking that wishes to be far more, ridiculously more, than it is.

References

Augustine. *City of God*, translated by Henry Bettenson. London: Penguin Books, 1984.

Bell, David. "On the Translation of Frege's *Bedeutung*." *Analysis* 40 (1980), 191–195.

Bencivenga, Ermanno. *Kant's Copernican Revolution*. New York: Oxford University Press, 1987.

Bencivenga, Ermanno. *Logic and Other Nonsense: The Case of Anselm and His God*. Princeton NJ: Princeton University Press, 1993.

Bencivenga, Ermanno. *Ethics Vindicated: Kant's Transcendental Legitimation of Moral Discourse*. New York: Oxford University Press, 2007.

Bencivenga, Ermanno. "Kripke's Critique of Materialism Debunked." *Epoché* 75 (2024). (2024a)

Bencivenga, Ermanno. "On Identity, Necessary and Contingent. Or: How the Precision of a Formal Language Can Be Fool's Gold." *Epoché* 73 (2024). (2024b)

Curley, Edwin. *Descartes against the Skeptics*. Cambridge MA: Harvard University Press, 1978.

Descartes, René. *The Philosophical Writings of Descartes*, in three volumes, translated by John Cottingham, Robert Stoothoff, Dugald Murdoch, and Anthony Kenny. Cambridge UK: Cambridge University Press, 1985/1991.

Dummett, Michael. *Frege: Philosophy of Language*. New York: Harper & Row, 1973.

Euclid. *The Thirteen Books of Euclid's Elements*, edited by Thomas Little Heath. Mineola NY: Dover, 1956.

Frege, Gottlob. *Aritmetica e logica*, translated by Ludovico Geymonat. Torino: Einaudi, 1948.

Frege, Gottlob. *Translations from the Philosophical Writings of Gottlob Frege*, by Peter Thomas Geach and Max Black. Oxford: Blackwell, 1952.

Frege, Gottlob. "The Thought," translated by Anthony and Marcelle Quinton. *Mind* 65 (1956), 289–311.

Frege, Gottlob. *Posthumous Writings*, translated by Peter Long and Roger M. White. Oxford: Blackwell, 1979.

Frege, Gottlob. *Philosophical and Mathematical Correspondence*, abridged for the English edition by Brian Francis McGuinness, translated by Hans Kaal. Oxford: Blackwell, 1980.

[30] Empirical evidence of this embarrassing misapprehension, by the actors in the social-media pathetic drama, of the very nature of what game they are playing, is their periodical transhumance from one social networking service to a more "secure" one. For those who have (transcendental) eyes to see, it is a phenomenon helpfully revealing how vulnerable their "privacy" is.

Freud, Sigmund. *The Standard Edition of the Complete Psychological Works of Sigmund Freud*, in twenty-four volumes, edited by James Strachey. London: The Hogarth Press, 1953/1974.

Gettier, Edmund Lee. "Is Justified True Belief Knowledge?" *Analysis* 23 (1963), 121–123.

Hintikka, Jaakko. "*Cogito, Ergo Sum*: Inference or Performance?" *The Philosophical Review* 71 (1962), 3–32.

Hodges, Andrew. *Alan Turing: The Enigma*. New York: Simon & Schuster, 1983.

Hume, David. *A Treatise of Human Nature*, edited by Lewis Amherst Selby-Bigge, second edition revised by Peter Harold Nidditch. Oxford: Clarendon Press, 1978.

Long, Peter, and Roger M. White. "On the Translation of Frege's *Bedeutung*: A Reply to Dr Bell." *Analysis* 40 (1980), 196–202.

Lorenz, Konrad. *On Aggression*, translated by Marjorie Kerr Wilson. New York: Harper, 1974.

Mendelsohn, Richard L. "Diary: Written by Professor Dr Gottlob Frege in the Time from 10 March to 9 April 1924," translated by Mendelsohn, edited and with commentary by Gottfried Gabriel and Wolfgang Kienzler. *Inquiry* 39 (1996), 303–342.

Metzger, Bruce Manning, and Roland Edmund Murphy (eds.). *The New Oxford Annotated Bible*. New York: Oxford University Press, 1991.

Plato. *Theaetetus*, translated by M. J. Levett. Indianapolis IN: Hackett, 1992.

Poe, Edgar Allan. *The Unabridged Edgar Allan Poe*, edited by Tam Mossman. Philadelphia: Running Press, 1983.

Ramsey, Frank Plumpton. *Foundations of Mathematics and Other Logical Essays*, edited by Richard Bevan Braithwaite. London: Harcourt Brace, 1931.

Russell, Bertrand Arthur William. "On Denoting." *Mind* 14 (1905), 479–493.

Russell, Bertrand Arthur William. *Introduction to Mathematical Philosophy*. London: George Allen & Unwin, 1919.

Chapter 6
Consciousness

Abstract On both sides of the analytic/continental divide, consciousness has enjoyed an exalted status, foundational for a world of its own or even for the world, period. On the contrary, I have a deflationary view of it: for me, a behavior X (and its subject) is (are) conscious/aware of another behavior Y (which may be identical with X) when it contains traces of Y—when Y is present, however slightly, within X—and X shows signs of reacting to it. Therefore, consciousness is all over: in animate and inanimate things, in plants, animals, and machines, as well as in human beings. And it goes with varying degrees of attention: from one that is focused to one that is peripheral, to one that is even blocked by forces internal to a subject. The Freudian distinction between what is conscious, preconscious, and unconscious finds its place here; but note that what is unconscious in this sense is still conscious in mine—it still has effect on one's behavior. Self-consciousness is consciousness of oneself, and its function is that of calling to attention all of a subject's resources in a situation that is perceived as dangerous. As such, it has none of the features Descartes attributes to it: it is not transparent, veridical, or privileged, since someone else can have better access to the reality of my situation than I do; and it is not continuous—it lasts as long as the situation is perceived as dangerous, and then fades. The delusion of a Cartesian (self-)consciousness is instigated by public powers, which feel threatened by the private sphere and hence insinuate there their own agent, ready to report on anything subversive going on (as the guard in Bentham's panopticon) and, to seal the deal, identified by the subject being spied on with its own most intimate nature. The duality consciousness/conscience hides the controlling role exercised by (self-)consciousness (apparent in Romance languages, where no such duality exists); though this role may turn out to have positive consequences if it works not in favor of oppressive political or social regimes but in favor of (Kantian, rational) ethics.

Keywords Consciousness · Conscience · Unconscious · Freud · Control · Panopticon · Ethics

Even among authors who are sympathetic to the line I have been developing here, there is frequent use of the word "consciousness" in talking about these positive

E. Bencivenga, *The Philosophy of (Artificial) Intelligence*,
https://doi.org/10.1007/978-3-032-09347-9_6

developments. Take, as an example, Peter Godfrey-Smith: his good book, mentioned in Chapter 2, promises in its title to be talking about the origins of consciousness. I would have used the word "intelligence" instead (that is the title of my chapter), or maybe "play," or "subjectivity"; "consciousness" I would have discarded.[1] And such examples could be multiplied at will.[2] For consciousness, or self-consciousness, is regarded (in non-theological discourse) as the defining property of humans; and merely suggesting (as Godfrey-Smith does) that it could be found not just in a non-human animal but even in one, like the octopus, not belonging among the mammals or the vertebrates, looks iconoclastic enough, and enough of an advertisement for one's work.

The originator of this view is, again, Descartes. (And again, it would rather be Augustine; but I will stay within the confines of the popular conversation.) I said that he finds his Archimedean fulcrum with which to raise the world in a direct contact with himself; now I can add that what establishes that contact is consciousness. It is not just that I, as a thinking thing, doubt, understand, affirm, deny, am willing or unwilling, imagine, and have sensory perceptions; it is also that I *know* I am doing all of that—and all else there is to know about me. My consciousness bears testimony to everything that goes on inside of me (which is the same as me—my outside, my body, is not me; it is just mine): it reports on it in a transparent, veridical, privileged, and continuous manner. Transparent: nothing is sheltered from it; I cannot hide from myself. Veridical: what it reports is true; I cannot lie to myself. (Which creates the paradox of self-deception, endemic for a Cartesian: for him, in order to deceive myself, I would have to both be deceived and be conscious of the deception, hence both know and not know one and the same thing.) Privileged: no one else has access to it; if anyone else wants to know what is going on in me, her only option is asking me (and her only hope is that my answer not be a lie). Continuous: it never stops reporting; I cannot lose sight of myself.

Kant did his best to sink this view by arguing, in the Refutation of Idealism, that consciousness is not a form of knowledge:

> Of course, the representation **I am**, which expresses the consciousness that can accompany all thinking,[3] is that which immediately includes the existence of a subject in itself, but not yet any **cognition** of it, thus not empirical cognition, i.e., experience; for to that there belongs, besides the thought of something existing, intuition, and in this case inner intuition, i.e., time, in regard to which the subject must be determined, for which outer objects are

---

[1] A personal anecdote: my Montaigne book (1990), referred to above, I had originally planned to title *The Discipline of Self-Consciousness*. It was in the course of writing it that I realized how wrong that title would have been and moved to the more appropriate one.

[2] I will cite only one more of them, from another sympathetic author: Dennett (1991).

[3] Notice that he says "*can* accompany" (italics added), thus implying that not all thinking is conscious, and that the I is not always directly involved in it. As I point out in my (2007), 136, in a letter to Jacob Beck of December 4, 1792, commenting on Beck's account of critical philosophy, he makes the following correction: "Instead of [...] 'The *I think* must accompany all the representations in the synthesis' 'must *be capable of* accompanying.'" This will become clearer when I explain that the "consciousness" mentioned here is self-consciousness.

> absolutely requisite, so that inner experience itself is only mediate and possible only through outer experience.[4]

Consciousness, for him, does not determine *what* it is that I am conscious of: does not provide any objective identification for it and does not distinguish it from anything it is *not*. For that to happen, consciousness must be projected onto *space*time and the subject of consciousness must find its place among *outer* objects. A fine point to make, within the purview of the Kantian articulation of objectivity (and objective knowledge) I have already provided room for in my framework. But one that did not toll the bell for Cartesians, who went through a major revival with Husserl (the author, among other things, of *Cartesian Meditations*[5]) and his phenomenology: the misguided attempt at analyzing objectively the nature of subjective, conscious experience (for a misunderstanding of which, I said, Searle was then responsible). The enduring popularity of this project has even been able to cross the otherwise sharp boundary between analytic and "Continental" philosophy, realizing a rare, unlikely concurrence between the two. Within the latter, Sartre regards the triumph of phenomenology as a *fait accompli*, and begins his *Being and Nothingness* with what looks like a straightforward, matter-of-fact statement: philosophy has at long last concluded that its only object of study is the phenomenon—what shows up in consciousness.

> Modern thought has realized considerable progress by reducing the existent to the series of appearances which manifest it. Its aim was to overcome a certain number of dualisms which have embarrassed philosophy and to replace them by the monism of the phenomenon.[6]

Within the former, the overwhelming presence of Cartesianism and the attendant philosophy of mind (which is, in effect, a philosophy of consciousness) has already been noted; and, if you want an explicit connection with Husserl, ask Searle.[7]

My view of consciousness is way more deflationary than that (but see below). It begins with a verbal move: I treat "conscious" and its cognates (here) as synonymous of "aware" and its cognates, in which the opposition is not as invested as in the former. To me, they are just a Latinate and a Germanic vehicle, respectively, for the same meaning, as are, say, "odor" and "smell." And I set a very low bar for consciousness/awareness: a behavior X (and its subject) is (are) conscious/aware of another behavior Y (which may be identical with X) when it contains traces of Y—when Y is present, however slightly, within X—and X shows signs of reacting to it. A mouse is conscious of a snake in the vicinity; the snake is conscious that the mouse is there; an immune

[4] Kant (1998a), 328 (B277).

[5] Husserl (1977).

[6] Sartre (1992), 3. Notice that this book's subtitle is *A Phenomenological Essay on Ontology*: according to it, its topic is being but its method of inquiry is phenomenology; as in Descartes, being is to be reached starting from the phenomena of consciousness.

[7] For a punctilious effort of disentangling Searle from too many dangerous liaisons with Husserl, and through him with Continental philosophy (conveniently acronymed as CP), see Mulligan (2003). (The reverent companion tome where this piece appears was published by a highly distinguished press before Searle had his emeritus status revoked by UC Berkeley for countless charges of sexual harassment. Talk of dangerous liaisons.)

system is conscious that its body is being attacked by a virus; the virus is conscious of the body's immune-system defenses and does its best to get around them; a geyser is conscious of coming to the end of its periodical spurt and a dishwasher is conscious of coming to the end of its cycle; a car is conscious of climbing a steep hill; the country is conscious that an economic crisis is impending. As I sit in a doctor's waiting room, I am simultaneously conscious of a lot of things: the book I am reading, the number of other patients waiting, the call I am expecting from a nurse to go in, the paintings on the walls. (So there is no point in speculating about the origins of consciousness: consciousness has always been all over the place.[8])

To make this clearer, and stress its deflationary flavor, seeing, smelling, and tasting (for animals, or even plants), as well as using more gas and shifting to a lower gear (for the car climbing a steep hill) are simply modes of consciousness: consciousness is not added to them as a badge of honor to elevate their status somehow. That I am conscious of the paintings on the waiting-room walls simply means that I see them, not that I see them *and also* (as a separate item in experience) I am conscious of seeing them; that the snake is conscious of the mouse simply means that it smells the mouse; that the car is conscious of climbing a steep hill simply means that it's using more gas and shifting to a lower gear; that the country is conscious that an economic crisis is impending simply means that the media are moaning, the government is contemplating emergency measures, the currency is under stress.

When X is conscious of Y, Y can be the target of various degrees of attention on the part of X. In the waiting room, most of my attention is devoted to the book I am reading: that is what I am focusing on. I am conscious of the paintings on the walls, having seen them when I entered, but I am paying no attention to them. In between these two extremities (intense attention and no attention at all) is the attention devoted to the call I am expecting from a nurse: I am not focusing on it, but I am alert enough to it (tuned in enough to it) to be ready to react as soon as it comes (where the readiness is already a reaction to it), and instantly close the book that at the moment is monopolizing my attention. This state is close to what Freud calls "evenly-suspended attention" (another name for it is "floating attention"), which, he recommends, is the kind that must be applied during psychoanalytic sessions:

> The technique [...] consists simply in not directing one's notice to anything in particular and in maintaining the same "evenly-suspended attention" [...] in the face of all that one hears. In this way we spare ourselves a strain on our attention which could not in any case be kept up for several hours daily, and we avoid a danger which is inseparable from the exercise of deliberate attention. For as soon as anyone deliberately concentrates his attention to a certain degree, he begins to select from the material before him; one point will be fixed in his mind with particular clearness and some other will be correspondingly disregarded, and in making this selection he will be following his expectations or inclinations. This, however, is precisely what must not be done. In making the selection, if he follows his expectations he is in danger of never finding anything but what he already knows; and if he follows his inclinations he will certainly falsify what he may perceive.[9]

[8] Nor is there for the origins of self-consciousness: as I explain later, it too has been all over the place.

[9] "Recommendations to Physicians Practising Psycho-Analysis," in Freud (1953/1974), vol. XII, 111–112.

This recommendation goes along with such other remarks as the following one:

> We know […] that many procedures are carried out with quite particular certainty if they are not the object of a special high degree of attention.[10]

But such caution against attention must be taken under judgment. For there are cases where directing to something "a special high degree of attention" does not bug performance: it is actually what is needed for the performance to succeed. If I am threading a needle, or dressing a wound, I'd better pay a great deal of attention to what I am doing; not so when I am running down a flight of stairs and begin paying attention to the complicated combination of deeds my muscles and nerves and bones are going through—then I am likely to trip and fall. The latter is what in English is dubbed being, or becoming, self-conscious, and that is a highly significant term, for it is through this route that self-consciousness enters here.

A subject being or becoming self-conscious will direct attention, even concentrated attention, not just to some aspect of its behavior, which may be (I said) just what is needed, but to itself. And that will often be cause for embarrassment—as suggested by the example of running down the stairs. An actor who becomes self-conscious in the middle of a performance may forget his lines; a runner who becomes self-conscious in the course of a race may lose it; a lover who becomes self-conscious in the course of asking "the question" may fumble his words. But it's not always that bad; sometimes, again, being self-conscious is exactly what is needed. If I am finding my way in a minefield, it's better to pay minute attention to any amount of pressure exercised by my feet, or any slight contact I make with something that could be a trip wire; if I am defending myself against an assault, it's smart to be keenly aware of how I am positioned, and of whether any wall, or door, or sun in my own or the other's eyes, confers me a tactical advantage or disadvantage (the same applies to a mouse that becomes conscious of a snake in the vicinity, or to a plant attacked by a parasite). And this is so much better (more useful) the less I am used to minefields or assaults; someone who does it all the time might be able to negotiate these situations thoughtlessly (no less than elegantly).

When superb athletes like Michael Jordan do what I just said (negotiate a situation thoughtlessly and elegantly), people say that they are "playing unconscious." What does that mean? It certainly does not mean that Jordan is not conscious, aware, of the ball, of the hoop, or of the positions of his teammates and opponents (in fact, champions like him have an uncanny ability to know, at any time, where everyone is on the floor, even without looking—whence the "no-look pass"). What it means is that the self does not influence his behavior, does not disturb it as signaled earlier by Freud. He is playing not unconscious but un-self-conscious. Which is as good a time as any to bring some clarification on the general notion of *unconscious*.

Freud distinguishes between what is conscious, which is what I am currently paying attention to (the book I am reading in the waiting room), what is preconscious, which is what I am not currently paying attention to but could be paying attention to if I wanted (the paintings on the room's walls), and what is unconscious, which

[10] *Introductory Lectures on Psycho-Analysis*, in ibid., vol XV, 30.

is what I could not be paying attention to even if I wanted (the sexual or aggressive fantasies that may have brought me to that very waiting room to see a doctor). All of these contents are conscious in my sense, because all of them leave a trace in my behavior and elicit reactions from it; but for some of them the subject (the self) will not recognize them as part of its own behavior, hence as part of itself—as far as they are concerned, it will not be *self*-conscious: not conscious of their being (part of) itself. The "self" in "self-conscious" indicates a direct object; "self-conscious" means "conscious *of oneself*." So, if I don't recognize elements that belong to my behavior as part of it, I am not self-conscious relative to them, or more plainly (if less grammatically) not self-conscious of them.

This can happen in both a temporary and instrumental, or a permanent and largely counterproductive, manner. I may be Michael Jordan and forget all about myself during the game, so that I not be in the way as Freud signaled; I will recover contact with myself, and even with my acrobatics, later, when it's time for interviews and eulogy; right now, I go with the flow, I am at one with the flow, I am not an individual distinct from the flow, *I am not there.* Or I may be a pathological subject, who keeps some of its behavior forever in a secluded state, divorced from the rest of it, unacknowledged, and cannot reconnect with it if not by external help. What I thus rule out is still me, and it is still having an effect (mostly, doing damage) in me; the sexual or aggressive drives I block out are the ones making me sick. This is what Freud called repressing, and I said already that the destiny of what is repressed is to press back and eventually return: this is where permanent (or semi-permanent if we allow for the return of it, and for the difference external help can make) limits of self-consciousness are in a whole lot of cases.

Midway between the healthy and the pathological is what is reported by Kuhn (1970):

> Scientists then often speak of the "scales falling from the eyes" or of the "lightning flash" that "inundates" a previously obscure puzzle, enabling its components to be seen in a new way that for the first time permits its solution. On other occasions the relevant illumination comes in sleep. (122-123)

In such cases the subject continues to play (un-self-consciously) around a prickling conundrum and at some point the play finds a crack to open it (this is what play will do), but the subject, for whatever reason, does not want to admit that it is its own work; so the behavior is not pathological in the sense that its outcome is felicitous but it is in the sense that that it remains permanently un-self-conscious. The entire paraphernalia of divine inspiration, of prophetic dreams, and of angels announcing some profound truth to mortals find their place here: in the insane asylum's vestibule though not right in the middle of it.

Saying more about the un(-self-)conscious must wait for another day. Let us bring self-consciousness back. How can we capture any positive contribution of it? What does it do when, instead of being awkwardly in the way, it fulfills a beneficial function?

What the examples above have in common is that in all of them—both those that end up badly and those that look more promising—the subject finds itself in a dangerous situation. The danger may be coming from the outside: a mine may blow up and tear me to pieces; the guy assaulting me may get away with my wallet, and my life. Or it may be coming from the inside: from a conflict internal to the subject itself. There may be something in me (of which I am not self-conscious) that does not really want me to ask the question, or wants me to fail as an actor or a runner: something that sees those outcomes as threats, and wants to avoid them. The range of examples can be extended, by including some of a different tenor but ultimately falling into the same category. If I am a novice on the basketball floor aspiring to be the next Michael Jordan but still prey to all sorts of naïve mistakes, I'd better pay attention, fierce attention, to what I am doing. The time for going with the flow might come later; now I am well-advised to self-consciously care for every step—as there is a danger that every step could be a wrong one.

Aristotle says something similar in the *Nicomachean Ethics*:

> Experience with regard to particular facts is also thought to be courage [...]. [S]oldiers exhibit [...] [this quality] in the dangers of war; for there seem to be many empty alarms in war, of which these have had the most comprehensive experience; therefore they seem brave, because the others do not know the nature of the facts.[11]

The expert at war will most often enjoy life in the barracks or the field, and be forgetful of his own importunate presence as he enjoys himself, since he knows that most often there is nothing to worry about; the novice not only will, but should, always regard himself as in peril, since his lack of experience in fact makes him so. He should always pay close attention to where he is, what surrounds him, how he could defend himself from a sortie, where he could escape if he needed to—before all of those things, after adequate training, become automatic and consciousness of oneself doing them or expecting them is no longer needed.

So this is what the function of self-consciousness is: as a siren that sounds when the self is in jeopardy and resources should be mobilized for its protection, as a call to arms when the enemy is setting its siege towers to get over the city walls (here the self would be the city), as a warning that one more blunder in executing the game plan will get the self kicked out of camp. Being put at the center of attention, the self becomes the main issue and all available energy will be directed to shielding it. The mechanism can be misused, as is shown by the examples that turn out badly, in that the self can take itself to be in danger when it's not—when asking the question, or winning the race, would in fact be the best thing that can happen to it—; but this is what mechanism it is. The function of a hammer is hitting away, and that is not contradicted by bringing up hammers that hit someone's head—a hammer can be useful nonetheless. The function of self-consciousness is giving a heads-up, and that is not contradicted if it be a vacuous, or even detrimental, heads-up—self-consciousness can be useful nonetheless.

Before I proceed, a terminological note. I told a story (of consciousness) that referred to characters, relations among them, and facts involving them. In telling the story, I used certain words to designate the characters, the relations, and the facts. I

---

[11] Aristotle (1984), 1762 (1116b).

would be telling the same story if I did so in German, or (overcoming my ignorance) in Chinese, or if I did so in English but using different words for the characters, the relations, and the facts—as long as the new distribution of words was isomorphic to the old one. Different structures of words can be used for different purposes (words are not innocent); but, as long as those structures perfectly match one another, nothing changes in the story being told.

In other works of mine,[12] I told the same story by using different words. "Consciousness" there stood for what is here "self-consciousness," which entailed that there was no consciousness at all of what relation I had with the book I was reading in the waiting room, the paintings on the walls there, or (if I were Michael Jordan) the position of every player on the floor. Another, less loaded, word could be used for that—perhaps "awareness," thus decoupling the synonymy I have adopted here. The reason is that those works meant to forcefully counter the Cartesian (phenomenological, mentalist) tradition,[13] where the archetype of consciousness, of what *they* call "consciousness," is the *cogito*, that is, self-consciousness; and I had to prove that the mental world built around the *cogito* is nearly empty (certainly nothing stays there for long), and no ontology can be reconstructed (pace Sartre and many others) on that scanty basis. So I went for broke: I had, after all, expressions like "playing unconscious" on my side and could consistently argue that most of what we do—when we act, when we read and write, and even when we think—is playing unconscious. Here I am more relaxed: I said that I am not going to bother mounting a concerted raid on the Cartesian hogwash, so I can allow people to be conscious of their surroundings, as long as it's clear that none of that will give them what the Cartesians hope and need.

Let us get to making that clear, then. Self-consciousness is not transparent: to the extent that a self is pathological, any number of things in it are opaque to it; even when it turns a thorough, insightful glance upon itself, they will not appear. Again, Freud can be called upon here for evidence. In his 1917 paper "A Difficulty in the Path of Psycho-Analysis,"[14] after mentioning the two "severe blows" (139) science has dealt to human narcissism by Copernicus dislodging humans from the center of the universe and Darwin denying the "gulf" (140) humans had placed between themselves and other animal species, he comes to the third one, and "probably the most wounding" (141), dealt by psychoanalysis.

> You feel sure that you are informed of all that goes on in your mind if it is of any importance at all, because in that case, you believe, your consciousness gives you news of it. And if you have had no information of something in your mind you confidently assume that it does not exist there. Indeed, you go so far as to regard what is "mental" as identical with what is "conscious"—that is, with what is known to you [...]. Come, let yourself be taught something on this one point! What is in your mind does not coincide with what you are conscious of; whether something is going on in your mind and whether you hear of it, are two different things. (142-143)

[12] See for example my (2003, 2023).

[13] The subtitle of my (2023) even says explicitly that my "metaphysical meditations" are set *against* Descartes (and on behalf of the pregnant dog kicked by Malebranche).

[14] Freud (1953/1974), vol. XVII, 137–144.

Freud's contribution is relevant to self-consciousness being veridical, too, since much of what it reports is mistaken: the result of elaborate forms of self-deception (yes! self-deception, since here the self is far from knowing everything that it is and does, hence it can be fooled by something it unknowingly does) rather than of self-enlightenment. And it is relevant to the privileged relation self-consciousness would establish between the self and its own contents (drives, conflicts, defenses, ...): often what is opaque to me about myself, what of it I even would inflexibly deny (denial is a form of defense[15]) is perfectly (transparently!) visible to an outside observer (who might then be able to provide external help).

As for continuity, it is out of the question. Self-consciousness arises quickly when danger, real or unreal, is on the horizon, and disappears just as quickly when it dissipates. Rehearsing an example I have used before (and rephrasing it in the language I have chosen here), say that I am driving on a lonely country road. I am conscious of the road, which I manage with great skill, never missing the right trajectory in a turn, the right speed at change of altitude, the right time to pass a slow truck; and I am conscious of the landscape, that I find charming, of the music the radio is playing, and of some fun memories I am indulging in—of when I was driving along the same road with someone dear to me. Then, suddenly, I hear a whistling sound coming from the car, and I wake up. I become conscious that *I* need to worry about what is happening, as I may be in danger: not grave danger, but danger nonetheless. The road *is* lonely; I haven't seen a gas station for miles; I am not sure my cellular phone will have reception; there are no comfortable spots where I could stop and wait for assistance (assuming I could call for assistance). If I had a mechanical breakdown here (once I had one in the middle of Death Valley), I could be stranded for hours, and I might have to trust a total stranger to give me a ride. This lasts for a while, during which time I am paying intense attention to myself, to my situation, and to what I should be doing. Then the noise, as it came, vanishes. I occupy myself some more in trying to decide what caused it. Perhaps, despite my "great skill," I have overused the brakes on this hilly terrain; perhaps I should have them checked as soon as a gas station turns up that looks as if there might be someone working there. But soon everything of the sort is forgotten; the self goes the same way as the potential danger; I am back to being conscious of the road, of the landscape, of the music, of those memories—to not being self-conscious at all.

Sirens (thank heavens!) don't sound all the time, or they would deafen us; nor do calls to arms or warnings.[16] Self-consciousness is (in healthy conditions) as intermittent as they are. It surfaces when there is a point to it, right or wrong as it might be, and goes underground as soon as that point is no longer relevant. There is no continuity to it; it may be the least continuous of our functions. When we are awake, we are continuously awake, and continuously conscious of a bunch of things; we are continuously in control of various physical abilities, and of the ability to speak our

[15] See Anna Freud (1966).

[16] Hyperconsciousness, often accompanied by anxiety, is not itself a psychological disorder but can be a symptom of one (for example, of the obsessive-compulsive disorder). We are about to see who, or what, a hyperconscious person is a victim of.

native language; we are continuously upbeat when we are upbeat and continuously pained when we are pained; but we are only sporadically, episodically, inextensively in the kind of touch with ourselves that is administered by self-consciousness.

Says Aristotle, in Book VII of the *Nicomachean Ethics*:

> [W]e should state not only the truth, but also the cause of error—for this contributes towards producing conviction, since when a reasonable explanation is given of why the false view appears true, this tends to produce belief in the true view.[17]

Putting the Cartesian view of self-consciousness next to mine may give an impression of "he says, she says": one view may feel as whimsical and unwarranted as the other. It strengthens one's position appreciably if one can explain why the other party should have arrived at a view as misdirected and rotten as it did. If self-consciousness is *not* (in the least) transparent, veridical, privileged, and continuous, why should those people claim that it is, indeed build an entire philosophy (of mind, of the phenomenon, of the world) on it?

It is again, initially, a question of words. "Consciousness" comes from the Latin "*conscientia*," hence from the Latin verb "*conscire*": *scire cum*, to know together. Originally, it referred to knowledge shared with another, though the phrases "*sibi conscire*," "*sibi conscius*" ("to know with oneself," "knowledgeable with oneself") were fairly common. They often had a moral tone: in his first epistle, addressed to his patron Maecenas, Horace writes: "*hic murus aeneus esto nil conscire sibi, nulla pallescere culpa*" (vv. 60–61; "be this our wall of bronze, to have no guilt at heart, no wrongdoing to turns us pale").[18] But sometimes they are not, and have a more neutral meaning, as in this judgment of Seneca by Quintilian:

> [M]y aim was not to ban his [*sc.* Seneca's] reading altogether, but to prevent his being preferred to authors superior to himself, but whom he was never tired of disparaging; for, being conscious [*sibi conscius*] of the fact that his own style was very different from theirs, he was afraid that he fail to please those who admired them.[19]

Christian authors strongly endorsed the moral sense, starting with Paul of Tarsus, who says, in his Epistle to Romans:

> When Gentiles, who do not possess the law, do instinctively what the law requires, these, though not having the law, are a law to themselves. They show that what the law requires is written on their hearts, to which their own conscience also bears witness; and their conflicting thoughts will accuse or perhaps excuse them on the day when [...] God [...] will judge the secret thoughts of all.[20]

The word translated here as "conscience" (about which more is coming) is "*conscientia*" in the Latin version of the epistle, and the corresponding word

---

[17] Aristotle (1984), 1824 (1154a).

[18] Horace (1962), 254–255.

[19] Quintilian (1968), 71–73. It may be worth mentioning that Horace and Quintilian were less than a century apart, and both well within the age of classical Latin.

[20] 2: 14–16, in Metzger and Murphy (1991), NT210-211.

"*syneídesis*" in Paul's Greek.[21] Augustine picks up this quote several times, for example in *Enarratio in Psalmum* 5: 13,[22] where, speaking of those who persevere in sin, he blasts:

> Let them fall by their own thoughts [...]; let them fall as their own thoughts accuse them and their own conscience [*conscientia*] bears witness against them, as the apostle says: "whether their thoughts accuse or defend them, when God's just judgment is revealed" (Rom 2: 15-16).

Aquinas goes the same way, if more dispassionately:

> [C]onscience [*conscientia*] is said to accuse or cause remorse, when that which has been done is found to be out of harmony with the knowledge according to which it is examined; or to defend or excuse, when that which has been done is found to have proceeded according to the form of the knowledge.[23]

Moving to modern times, we have Descartes state, in the Replies to the First Objections, that "*potui [...] pro certo affirmare nihil in me, cujus nullo modo sim conscious, esse posse*" ("I could assert with certainty that nothing could be in me of which I was not in any way conscious"),[24] and it belongs to common lore that the word "consciousness," in this specific Cartesian sense of self-consciousness, entered philosophical English with Locke's *Essay*, where consciousness is defined as "the perception of what passes in a Man's own mind."[25] Some twenty years later Samuel Clarke wrote, in one of his letters to Anthony Collins: "*Consciousness*, in the most strict and exact sense of the word, signifies [...] the *reflex act by which* I *know that* I *think, and that my thought and actions are my own and not another's*."[26] And Thomas Reid concurred: "Consciousness is a word used by Philosophers, to signify that immediate knowledge which we have of our present thoughts and purposes, and, in general, of all our present operations of our minds".[27] But one thing that is less insisted on is the following: whereas in Romance languages, which stay close to Latin, a single word is used for both the moral and the neutral sense ("*conscience*" in French, "*conciencia*" in Spanish, "*consciência*" in Portuguese, "*coscienza*" in Italian, "*conştiinţă*" in Romanian), when that word, or some corresponding one,

---

[21] This word is constructed in the same way as "*con-scientia*," with "*syn*" meaning "*cum*" and "*eídesis*" connected with the verb "*eidénai*" (to see, to know), it has the same original meaning of *knowing together*, and the same derivative meanings, moral and neutral, as "*conscientia*" (though in the New Testament the moral meaning is predominant). In a text to be mentioned shortly, Aquinas, after quoting Jerome, says "it seems that conscience and synderesis are the same thing"; and apparently "*syndéresis*," which became popular among Medieval authors, was the result of a corruption from "*syneídesis*," introduced precisely through copying Jerome texts.

[22] See it at https://www.augustinus.it/latino/esposizioni_salmi/esposizione_salmo_005_testo.htm; translation mine.

[23] *Quaestiones disputatae de Veritate* 17:1. See it at https://isidore.co/aquinas/QDdeVer17.htm.

[24] Descartes (1904), 107.

[25] Locke (1975), 115.

[26] Uzgalis (2011), 90.

[27] Reid (1785), 17. Though notice the persistence of the original Latin sense in this passage from Hobbes (1985): "When two, or more men, know of one and the same fact, they are said to be CONSCIOUS of it one to another," 131–132.

enters a Germanic tongue, it divides up. Starting with Locke, English reserves "consciousness" for the neutral, or if you will purely cognitive, sense, whereas the older word "conscience" (which is closer to the Latin, and to the ones in use in Romance languages) remains attached to the moral one. Thus we see Hamlet say that "conscience does make cowards of us all"[28] (vividly anticipating what my position on the matter is going to be) and, for an accurate definition, here is Merbecke (1581):

> THe conscience verilie is the knowledge, iudgement, & reason of a man, whereby euerie man in himselfe, and in his owne minde, being made priuie to euerie thing, y[t] he either hath committed or not committed, doe either condemne or acquite himself.

Locke himself retains both words, with their distinct senses:

> *Conscience* [...] is nothing else, but our own Opinion or Judgment of the Moral Rectitude or Pravity of our own Actions.[29]

And, as for Kant's German, it is "*Bewußtsein*" for the consciousness discussed in the Refutation of Idealism but "*Gewissen*" for what is discussed in such passages as the following, from *The Metaphysics of Morals*:

> [C]onscience is practical reason holding man's duty before him for his acquittal or condemnation in every case that comes under a law.[30]

Now the essence of what I am going to say about Cartesian consciousness is that its ambiguous relation to conscience reveals the bad faith (the bad conscience) expressed in it: it reveals that, despite its alleged neutral, anodyne stance, Cartesian consciousness has a judgmental, persecutory, hampering function. To introduce it, I will bring up a work by Jeremy Bentham, philosopher, jurist, and social reformer, the founder of Utilitarianism, and an early proponent of the rights of non-human animals,[31] who, however, got at times a little carried away in his plans for social reform. I am referring specifically to a writing of his called *Panopticon; or, The Inspection House*, written as a series of letters in 1787 to "a friend in England" from Crecheff in White Russia (modern-day Krichev, in Belarus).[32] Here Bentham proposes a new structure for prisons, whose idea he attributes to his brother, and which he believes can be of much larger application:

---

[28] Shakespeare (1975), 1088.

[29] Locke (1975), 70.

[30] In Kant (1999), 529.

[31] In his (1907) he says: "It may come one day to be recognized, that the number of the legs, the villosity of the skin, or the termination of the *os sacrum*, are reasons equally insufficient for abandoning a sensitive being to [...] [the caprice of a tormentor]. [...] [T]he question is not, Can they *reason*? nor, Can they *talk?* but, Can they *suffer*?" (311n). I feel I should mention at least this, in light of how critical of a Bentham proposal I am going to be. I add that, in years subsequent to the text I am about to focus on, well over fifty prisons were built in England according to Bentham's model.

[32] Bentham (1791). Foucault (1977) argued that the panopticon represented a form of disciplinary social control that was to have a large influence in the contemporary world. I agree, but extend the point from structures of political administration to their significance in terms of subjectivity.

> [I]t will be found applicable, I think, without exception to all establishments whatsoever, in which within a space not too large or commanded by buildings, a number of persons are meant to be kept under inspection. No matter how different, or even opposite the purpose: whether it be, that of *punishing the incorrigible, guarding the insane, reforming the vicious, confining the suspected, employing the idle, maintaining the helpless, curing the sick*, or *training the rising race* in the path of *education*: in a word whether it be applied to the purposes of *perpetual prisons* in the room of death, or *prisons for confinement* before trial, or *penitentiary-houses*, or *houses of correction*, or *work-houses*, or *manufactories*, or *madhouses*, or *hospitals*, or *schools*. (2-3)

Here is what the structure looks like:

> The building is circular. The apartments of the prisoners [but remember: these could also be patients, workers, or students] occupy the circumference. You may call them, if you please, the *Cells*. These *Cells* are divided from one another, and the prisoners by that means secluded from all communication with each other [...]. The apartment of the Inspector occupies the center; you may call it if you please the *Inspector's Lodge*. [...] The inner circumference of the Cell is formed by an iron *grating*, so light as not to screen any part of the Cell from the Inspector's view. [...] To the windows of the Lodge there are *blinds*, as high up as the eyes of the prisoners in their Cells can, by any means they can employ, be made to reach. [...] Small *lamps*, in the outside of each window of the Lodge, backed by a reflector, to throw the light into the corresponding Cells, would extend to the night the security of the day. (5-8)

That is: the inmates would be exposed to the eyes of the inspector night and day, but they would never know whether the inspector is actually looking at them ("there are *blinds*"). In Bentham's words: "the most effectual contrivances" are in place for the inspector "*seeing without being seen*" (23). It is perhaps "the most important point," he adds, "that the persons to be inspected should always feel themselves as if under inspection, at least as standing a great chance of being so" (24). How important this point is gets expressed in the clearest—and to us most relevant—way when this deranged scheme is extended to schools, and we are told:

> All play, all chattering, in short all distraction of every kind, is effectually banished by the central *and covered* situation of the master. (122; italics added)

People constantly under scrutiny will not transgress; but it may be next to practically impossible to maintain active scrutiny on a given population all the time; so we move from a collective to a distributive (and partially fictional) understanding of that "constantly." It is not necessary that *the whole* population be inspected *at all times*; it is enough that *any member of it* believe *at any one time* that they might be—and then all it takes is that they never know if they are under scrutiny or not. That trick will make them invariably inclined to comply. To stress the analogy with self-consciousness (with the ideology of self-consciousness) I intend to bring up, the trick consists of convincing the inmates that the inspector has a transparent, veridical, privileged (there are no other guards, and the inmates have no perception of each other), and continuous access to them, whereas for all they know he might be spending his time reading the newspaper or entertaining women friends, or there might even be no inspector at all behind the blinds.

Let me bring up the analogy, then. Mention of play in the last quote is revealing. A child who spends time in school playing, instead of adhering to harsh disciplinary rules, will be a potential agent of subversion for those rules. Back in the days when I went to school, it would have been hit hard, physically, by the teacher in order to (attempt to) bring it into line (that never succeeded with me); today physical punishment is banned but, if the child strands too far from the line, it might receive a diagnosis of ADHD and be doped for good. So it will hide its play, indulge in amusing itself when it's sure no one (especially the teacher) is looking. And the same will be true for the residents of all the other social institutions where Bentham plans to apply his brother's invention: people who don't fall under the conventional definition of sanity, workers whose every minute is exploited for profit, vagrants who are being "educated" to a "productive" life, addicts who prefer their own drug of choice to the stupor induced by late capitalism will always be looking for private circumstances, from which others are excluded, to practice their own deviant play—and who knows what might issue from it? maybe an alternative to late capitalism. Therefore, private space itself must be denied: no escape must be left from constant inspection. These days, surveillance mechanisms are provided by the digital invasion of that space, primarily in the form of "social" media; when that technology was not available, entrenched powers (first and foremost the Church, curiously not listed by Bentham among his possible clients—because it was already in that line of business?) found a more subtle means. They insinuated an external agent inside the subject itself, and convinced the subject that it represented its most intimate contact with itself.

It is, as with everything metaphysical, a political operation. There is a war going on at all times between the impertinence of play and the stability of those subjects and behaviors that have chosen steadiness, and may even want to use play, by themselves or others, to reinforce their bulwarks. Increasingly sophisticated weapons are used, by both sides, to improve their chances. Jules Verne (1874) tells humorously of one such exchange between Impey Barbicane, a maker of shells, and Captain Nicholl, a maker of plates:

> Now if Barbicane was a great founder of shot, Nicholl was a great forger of plates; the one cast night and day at Baltimore, the other forged day and night at Philadelphia. As soon as ever Barbicane invented a new shot, Nicholl invented a new plate, each followed a current of ideas essentially opposed to the other. [...] At the last experiment the cylindro-conical projectiles of Barbicane stuck like so many pins in the Nicholl plates. On that day the Philadelphian iron-forger then believed himself victorious [...]; but when the other afterwards substituted for conical shot simple 600lb. shells, at very moderate velocity, the captain was obliged to give in. In fact, these projectiles knocked his best metal plate to shivers. Matters were at this stage, and victory seemed to rest with the shot. (50)

Same thing here; let us review the main steps of this confrontation, as we have encountered them so far. Play is instructive and innovative but is also risky: a child playing can get itself electrocuted, or burn down the house; a medical researcher playing with a virus can cause the next plague. Therefore, all sorts of protections will be erected around it, and within it: thick barriers to avoid its spreading to vulnerable environments; substitute objects (toys) that are far less dangerous (but from which one also learns far less); confinement of play into the safer arena of language, where the substitutes for objects are words and phrases. All of these means are public, though, and the threat of someone being inspired by a pantomime or a story to try a real thing, or stealing the idea expressed by the pantomime or the story before its originator is able to develop it, and possibly profit from it, is still high; so confinement is made radical, by cutting all connections with the outside and inventing a private space which is only accessible to its own subject—and where private language becomes thought. It is a brittle space, always about to lose its integrity, always tempted by the prospect of relieving tension and going public; but, with all of its tenuousness, it is the best device contrived so far for keeping a subject's play under its own mastery—for indulging it in secret and publishing it (if at all) only at the most opportune time. What self-consciousness does is turn this precious device against itself, against its own function, and against the subject whose function it is. An evil eye is made part of privacy, to expose its transgressions and thus to "make cowards of us all." In early stages of this stratagem, the evil eye was an external, supernatural agent's, to whom nothing is hidden. (Does God see me when I go to the bathroom? would a child ask when first introduced to that concept.) Then comes the last, most effective wrinkle: the eye is one's own, indeed it is oneself—one has, for all intents and purposes, come to the point of policing oneself.

Here is Freud, from his *Introductory Lectures on Psycho-Analysis*:

> [T]here actually exists in the ego an agency which unceasingly observes, criticizes and compares, and in this way sets itself over against the other part of the ego. We believe, therefore, that the patient is betraying a truth to us which is not yet sufficiently appreciated when he complains that he is spied upon and observed at every step he takes and that every one of his thoughts is reported and criticized. [...] We know the self-observing agency as the ego-censor, the conscience [*Gewissen*]; it is this that exercises the dream-censorship during the night, from which the repressions of inadmissible wishful impulses proceed. When in delusions of observation it becomes split up, it reveals to us its origin from the influences of parents, educators and social environment—from an identification with some of these model figures.[33]

By not recognizing the relation between *Gewissen* and *Bewußtsein*, between conscience and consciousness, Freud is not able to disclose the culmination of this spy work: if it is true that conscience is society exercising censorship on the subject, it is also true that (self-)consciousness makes the subject identify with that very censorship, and exercise it on itself. Kant brings out this ambiguous situation when he says, for example,

> [A human being would] feel himself compelled by reason to acknowledge his judgment [that he not be adequate to a perfectly moral world] with complete impartiality, as if rendered by somebody else yet at the same time his own.[34]

This completes the second, Aristotelian part of what I set up to do in the present chapter. After explaining what consciousness and self-consciousness are, I have now explained why they are so often taken the wrong way: it is because our understanding of them is vitiated by an inimical operator, and a powerful one, which works, most often with great success, to distort the view we have of ourselves.

---

[33] Freud (1953/1974), vol. XVI, 428–429.

[34] Kant (1998b), 59.

A final note is in order, related to the last Kant quote and relevant to the topic of the next chapter. The arms race between the private and the public can go one step further, and once more twist things around: the nosy character the public has introduced into private space to expose its subversiveness and make it hostage to the powers that be can be turned by the subject into a tool for monitoring its own moral integrity. Instead of being made by it into a slave to abusive social or religious demands, the subject can use it as part of an effort of conforming to the rationality of the categorical imperative, if necessary by *countering* those demands—this is what, specifically, Kant is talking about. As I said before, morality will not be in the forefront of my discussion in this book; but it's good that it be at least in the margins of it, since it will come up now and again (most notably, at the very end), and its presence (as something preconsciously there) will enrich the fiber of the cloth being weaved.

References

Aristotle. *The Complete Works of Aristotle*, edited by Jonathan Barnes. Princeton NJ: Princeton University Press, 1984.

Bencivenga, Ermanno. *The Discipline of Subjectivity: An Essay on Montaigne*. Princeton NJ: Princeton University Press, 1990.

Bencivenga, Ermanno. *Dancing Souls*. Lanham MD: Lexington Books, 2003.

Bencivenga, Ermanno. *Ethics Vindicated: Kant's Transcendental Legitimation of Moral Discourse*. New York: Oxford University Press, 2007.

Bencivenga, Ermanno. *Meditazioni metafisiche. Contro Cartesio e in nome di una cagna gravida*. Napoli: Guida, 2023.

Bentham, Jeremy. *An Introduction to the Principles of Morals and Legislation*. Oxford: Clarendon Press, 1907.

Bentham, Jeremy. *Panopticon; or, The Inspection-House*. Printed in Dublin, reprinted in London 1791. At https://archive.org/details/panopticonorins00bentgoog/page/n2/mode/2up.

Dennett, Daniel. *Consciousness Explained*. Boston: Little, Brown, and Co., 1991.

Descartes, René. *Oeuvres*, edited by Charles Adam and Paul Tannery, vol. VII. Paris: Librairie Philosophique J. Vrin, 1904.

Foucault, Michel. *Discipline and Punish: The Birth of the Prison*, translated by Alan Sheridan. New York: Pantheon Books, 1977.

Freud, Sigmund. *The Standard Edition of the Complete Psychological Works of Sigmund Freud*, in twenty-four volumes, edited by James Strachey. London: The Hogarth Press, 1953/1974.

Freud, Anna. *The Ego and the Mechanisms of Defence*, revised edition. London: Karnac Books, 1966.

Hobbes, Thomas. *Leviathan*. London: Penguin Books, 1985.

Horace. *Satires, Epistles and Ars Poetica*, translated by Henry Rushton Fairclough. Cambridge MA: Harvard University Press, 1962.

Husserl, Edmund. *Cartesian Meditations*, translated by Dorion Cairns. Leiden NE. Martinus Nijhoff, 1977.

Kant, Immanuel. *Critique of Pure Reason*, translated by Paul Guyer and Allen William Wood. Cambridge UK: Cambridge University Press, 1998. (1998a)

Kant, Immanuel. *Religion and Rational Theology*, translated by Allen William Wood and George di Giovanni. Cambridge UK: Cambridge University Press, 1998. (1998b)

Kant, Immanuel. *Practical Philosophy*, translated by Mary Jeanne Gregor. Cambridge UK: Cambridge University Press, 1999.

Kuhn, Thomas Samuel. *The Structure of Scientific Revolutions*, second edition. Chicago: The University of Chicago Press, 1970.

Locke, John. *An Essay concerning Human Understanding*, edited by Peter Harold Nidditch. Oxford: Oxford University Press, 1975.

Merbecke, John. *A Booke of Notes and Common Places*. London: Thomas East, 1581. At https://quod.lib.umich.edu/e/eebo/A06863.0001.001/1:6.59.1?rgn=div3;view=fulltext.

Metzger, Bruce Manning, and Roland Edmund Murphy (eds.). *The New Oxford Annotated Bible*. New York: Oxford University Press, 1991.

Mulligan, Kevin. "Searle, Derrida and the Ends of Phenomenology," in *The Cambridge Companion to Searle*, edited by Barry Smith. Cambridge UK: Cambridge University Press, 2003, 261–286.

Quintilian. *Institutio oratoria*, vol. IV, translated by Harold Edgeworth Butler. Cambridge MA: Harvard University Press, 1968.

Reid, Thomas. *Essays on the Intellectual Powers of Man*. Edinburgh: John Bell, 1785.

Sartre, Jean-Paul. *Being and Nothingness: An Essay in Phenomenological Ontology*, translated by Hazel Estella Barnes. New York: Washington Square Press, 1992.

Shakespeare, William. *The Complete Works*. New York: Crown, 1975.

Uzgalis, William L. (ed.). *The Correspondence of Samuel Clarke and Anthony Collins, 1707-08*. Peterborough ON: Broadview, 2011.

Verne, Jules. *From the Earth to the Moon Direct in Ninety-Seven Hours and Twenty Minutes and a Trip Round It*, translated by Louis Mercier and Eleanor E. King. New York: Scribner, Armstrong & Company, 1874.

Chapter 7
Will

Abstract The hold that the will is supposed to exercise on some human behavior has always been feeble: though authors like Hume stated that we are free whenever we act in accordance with our intentions (that is, spontaneously), Kant has argued that such appeal to intentions is a wretched subterfuge—what we want or will is no less determined by previous factors, and ultimately factors well beyond the scope of our lives, than what else we do. As an alternative to such chicanery, he developed his own brand of compatibilism: a behavior is autonomous if, in addition to being determined by natural causes (which it always is), it is also rational, and hence reason in us can appropriate it and declare it a manifestation of itself. Despite the uninformed, tendentious readings "scholars" have offered for the so-called "formula of humanity," rationality cannot be limited within Kant's position to members of the species *homo sapiens*: animals, plants, and machines can also display it, and hence be attributed a free will (Kantian *Wille*). Turning now to the more modest display of intentions, in a Humean vein, they too cannot be limited to humans: Freud has provided ample evidence that our intentions (and even his own) typically escape the subject's consciousness and are best seen by an outside observer. So I can attribute intentions to a machine with as much legitimacy as I do to a human, and what that human judges (from the "inside") to be his intentions is as much conjectural and uncertain as my (external) hypotheses are.

Keywords Will · Determinism · Compatibilism · Spontaneity · Kant · Rationality · Intentions · Freud

In a work published (anonymously) in 1782 and entitled *De' doveri de' Principi neutrali verso i Principi guerreggianti, e di questi verso i neutrali* (*Of the Duties of Neutral Princes toward the Warring Ones, and of the Latter toward the Neutral Ones*), the Italian Enlightenment thinker Ferdinando Galiani (1728–1787) writes:

> As for myself, I say that *necessity* has no rights, nor can it generate any. It has privileges. But all its privileges it shares with ignorance, inadvertence, insanity, delirium, in sum with any state in which man operates without the freedom of acting, and without knowing what he's doing, or being able not to do what he does. Without free will there is no merit or demerit in

E. Bencivenga, *The Philosophy of (Artificial) Intelligence*,
https://doi.org/10.1007/978-3-032-09347-9_7

> actions. Man becomes a machine, and a machine carries no blame. But death given to a man by a machine is not for that reason just. You can only say that the machine has committed no crime.[1]

The necessity mentioned here is moral, not physical, physiological, or psychological: it is the necessity of fulfilling one's own duties toward oneself—duties which may become relevant when someone else threatens one's survival. But that condition is quickly assimilated with such physical, physiological, or psychological ones as insanity or ignorance, and the general sense is clear: unless a human has control of her actions—she knows what she is doing, and she is (physically and morally) able to do what she wants—her actions (assuming we can even call them that) are not deserving of praise or blame. Since a machine is never "able not to do what [...] [it] does," nothing a machine does should be praised or blamed.

Aristotle agrees that only when a human freely chooses (wills) what to do is it legitimate to pass a moral judgment on what she does. In the *Nicomachean Ethics* we read:

> Since excellence is concerned with passions and actions, and on voluntary passions and actions praise and blame are bestowed, on those that are involuntary forgiveness, and sometimes also pity, to distinguish the voluntary and the involuntary is presumably necessary for those who are studying excellence, and useful also for legislators with a view to the assigning both of honours and of punishments.[2]

He goes on to say that an action is involuntary when it is either forced or caused by ignorance, but gives a strict interpretation of what that means. "Caused by ignorance" is limited to not knowing some crucial detail of the circumstances in which one is acting (Oedipus does not know that the man he killed was his father, or that the woman he married is his mother—the excuse does not extend to being ignorant of such general laws as that homicide and incest are crimes, or to cases in which one is acting in ignorance which is, however, caused by oneself, as when one chooses to get drunk and lose all notion of what he's doing); most importantly, "being forced" is limited to not having physical control of one's limbs, and hence of one's movements, as when someone "were to be carried somewhere by a wind, or by men who had him in their power."[3] On the other hand, acting under duress, that is, under what many would regard as unsustainable psychological pressure, does not qualify:

> [I]f a tyrant were to order one to do something base, having one's parents and children in his power, and if one did the action they were to be saved, but otherwise would be put to death [...] the man acts voluntarily; for the principle that moves the instrumental parts of the body in such actions is in him, and the things of which the moving principle is in a man himself are in his power to do or not to do.[4]

---

[1] Galiani (1975), 671n; translation mine.

[2] Aristotle (1984), 1752 (1109b). Notice that the Greek word here translated as "excellence" ("*areté*") is more commonly translated as "virtue."

[3] Ibid. (1110a).

[4] Ibid.

The cases cited by Galiani as deprived of free will would seem to be captured by Aristotle's definition: most likely, an insane or delirious person (Agave, say, in Euripides' *Bacchae*) would seem to be ignorant of the specific circumstances in which she is acting (though we might debate on how responsible she is for getting delirious in the first place). And, presumably, a machine would be as incapable of voluntary action for Aristotle as for Galiani, since it would be as irresistibly "carried" by its program as a person might be by the wind. But the real philosophical challenge, in this neighborhood, comes from the authors who extend the "excuses" allowed by Aristotle and Galiani to the totality of human behavior, which they regard as unfailingly determined by forcing factors of all kinds—genetic makeup, drives, society, education—arranged in necessary causal sequences which extend far beyond anyone's life, and claim that a human is, as a consequence, *never* free to act, that she could never do otherwise than exactly what she does. This thesis of *determinism* has been a constant presence in the history of (Western) philosophy, from the Presocratic atomism of Leucippus and Democritus through the Stoics, Hobbes, Leibniz, Laplace, and William James all the way to the recent Sapolski (2023). For a vibrant formulation of it, see Nietzsche in *Beyond Good and Evil*:

> The longing for "freedom of the will" in the superlative metaphysical sense (which, unfortunately, still rules in the heads of the half-educated), the longing to bear the entire and ultimate responsibility for your actions yourself and to relieve God, world, ancestors, chance, and society of the burden—all this means nothing less than being that very *causa sui* and, with a courage greater than Münchhausen's, pulling yourself by the hair from the swamp of nothingness into existence. Suppose someone sees through the boorish naiveté of this famous concept of "free will."[5]

More relevantly for us, in the same eighteenth century when Galiani lived, Julien Offray de La Mettrie (1709–1751) drew from determinism the conclusion that there is no difference between humans and machines,[6] and made that the title of his most famous work, *Man as Machine* (*L'homme machine*).[7] Here is a typical passage:

> The human body is a machine which winds its own springs. It is the living image of perpetual movement. Nourishment keeps up the movements which fever excites. Without food, the soul pines away, goes mad, and dies exhausted. The soul is a taper whose light flares up the moment before it goes out. But nourish the body, pour into its veins life-giving juices and

[5] Nietzsche (2002), 21.

[6] Long before the current craze on AI, automata (machines that could reproduce human behaviors) were extremely popular in the eighteenth century. I will mention at least three that were built between 1768 and 1774 by the Swiss-born watchmaker Pierre Jaquet-Droz, his son Henri-Louis, and Jean-Frédérick Leschot. The Writer had a programmable mechanical memory and could write any text up to forty letters long; The Musician could play a custom-built organ by pressing its keys (no prerecorded music); The Draughtsman could draw four different images. The Jaquet-Droz automata can be seen at the art and history museum in Neuchâtel, Switzerland.

[7] It is relevant to our purposes here to point out that the Frenchman La Mettrie was forced to find refuge in Leiden when his views made him unpopular in his country, and eventually had to leave even relatively liberal Holland (where he had written *Man a Machine*) for further exile in Prussia, where he died. Crossing the boundary between humans and machines has always been frowned upon by a universal anthropocentric regime, which even invented (as I pointed out already) various gods in its image to provide a fake supernatural base to its tyranny.

> strong liquors, and then the soul grows strong like them, as if arming itself with a proud courage, and the soldier whom water would have made flee, grows bold and runs joyously to death to the sound of drums. Thus a hot drink sets into stormy movement the blood which a cold drink would have calmed.[8]

In the face of determinism, many philosophers (including some of those cited above as supporters of it) have tried to find a place for voluntary, free action by espousing some variant of *compatibilism* (the view that determinism and free will are compatible). A classic example is Hume, in the *Treatise*:

> Few are capable of distinguishing betwixt the liberty of *spontaneity*, as it is call'd in the schools, and the liberty of *indifference*; betwixt that which is oppos'd to violence, and that which means a negation of necessity and causes. The first is even the most common sense of the word; and as 'tis only this species of liberty, which it concerns us to preserve, our thoughts have been principally turn'd towards it, and have almost universally confounded it with the other.[9]

Spontaneity is what is denied when the wind carries me away, thus doing violence to me; indifference is what determinists deny, claiming that I could never do other than I did. Hume (somewhat inconsistently, as he was the most vocal and effective critic of necessary causal relations) thinks that determinists are right:

> We feel that our actions are subject to our will on most occasions, and imagine we feel that the will itself is subject to nothing [...]. But [...] we can never free ourselves from the bonds of necessity. We may imagine we feel a liberty within ourselves; but a spectator [so much for privileged access to one's mental life; more about it later] can commonly infer our actions from our motives and character [...]. Now this is the very essence of necessity, according to the foregoing doctrine [of determinism].[10]

We needn't worry, though: as long as we can "freely" move our limbs, as long as we do what we "want" to do without outside interference, that is all we should care about, and never mind if *wanting* to move limbs (no less than moving them) a certain way is itself a necessary consequence of previous causal factors.[11]

Kant judges this attitude a fraud:

> It is a wretched subterfuge [...] [to call free] the actions of the human being, although they are necessary by their determining grounds which preceded them in time, [...] because [...] [they] are caused from within, by representations produced by our own powers, whereby desires are evoked on occasion of circumstances and hence actions are produced at our own discretion [...]. Some still let themselves be put off by this subterfuge and so think they have solved, with a little quibbling about words, that difficult problem on the solution of which millennia have worked in vain and which can therefore hardly be found so completely on the surface.[12]

---

[8] La Mettrie (1912), 93.

[9] Hume (1978), 407–408.

[10] Ibid., 408–409.

[11] That we should not worry too much about facing the strongest and most obvious challenge of determinism is indicated by the very title of a recent contribution to the compatibilist literature: Dennett (1984).

[12] *Critique of Practical Reason*, in Kant (1999), 216–217.

And he proceeds to unfold his own brand of compatibilism, where an action, which is always determined by antecedent causal factors, is also free, and an expression of will (*Wille*), when it is rational, and hence manifests (in addition to the necessity of nature) the autonomy (self-legislation) of reason. I have accounted for this complex, and so far perversely misunderstood, doctrine elsewhere;[13] here I will add only one remark. For Kant, rational autonomy may show up in human behavior (a human may hope that it show up in his own), but only because humans are rational, not because rationality *coincides* with humanity:

> [W]e must not let ourselves think of wanting to derive the reality of [...] [the] principle [of duty] from *the special property of human nature*. For, duty is to be practical unconditional necessity of action and it must therefore hold for all rational beings [...] and only *because of this* can it also be a law for all human wills.[14]

The so-called (and itself often perversely misunderstood[15]) "formula of humanity" of the categorical imperative asserts: "*So act that you use humanity, whether in your own person or in the person of any other, always at the same time as an end, never merely as a means.*"[16] But that, consistenly with the previous quote, is not supposed to make any special provisions for featheless bipeds; in fact, a few lines before stating the formula, Kant says that its ground is "*rational nature exists as an end in itself*,"[17] and the formula is rephrased more than once to dispel any possible misunderstanding of it— "the human being and in general every rational being *exists* as an end in itself, *not merely as a means* to be used by this or that will at its discretion; instead he must in all his actions, whether directed to himself or also to other rational beings, always be regarded *at the same time as an end*;"[18] "so act with reference to every *rational being* (yourself and others) that in your maxim it holds at the same time as an end in itself."[19]

As it turns out, Kant thinks that humanity is the only example of rationality we encounter in experience:

> [W]e cannot imagine another suitable form for a rational being than the form of man.[20]

> [T]he problem of giving an account of the character of the human species is quite insoluble, because the problem could only be solved by comparing two species of rational beings on the basis of experience, but experience has not offered us a comparison between two species of rational beings.[21]

---

[13] In my (2007).

[14] *Groundwork of the Metaphysics of Morals*, in Kant (1999), 76.

[15] Those misunderstanding it should explain why Kant says "[t]he [...] three ways of representing the principle of morality [including, then, the formula of humanity] are at bottom only so many formulae of the very same law" (ibid., 85)—where the others make no mention of humanity. But (as there is no end to foolishness, or bad faith) I have heard people accuse Kant *of inconsistency* for this very reason that they made up.

[16] Ibid., 80

[17] Ibid., 79.

[18] Ibid.

[19] Ibid., 87 (italics added).

[20] Kant (1978), 62.

[21] Ibid., 238.

I believe that he was wrong in thus limiting the range of our experiencing rationality, and in a subsequent chapter I will explain why; but, however that goes, it remains true (which is the substance of my current remark) that, as far as Kant is concerned, nothing rules out a Martian, an animal, a plant, a society (of humans, animals, plants, or Martians), *or a machine* manifesting rationality in their behavior, and hence this behavior counting as voluntary and free. No necessary link exists between membership in the species *homo sapiens* and the presence of *Wille.*

The speculations suggested by this last remark are of enormous scope. As the postulates of practical reason ask us—not to believe, let alone affirm, in a theoretical sense, but—to be guided in our action by a commitment to the eventual triumph of reason, this commitment might involve seeing the triumph take place through a transcending of *human* rationality into some other, more vast and adequate form of it. I will return briefly to such overweening prospects at the end; here I abandon them, together with the whole issue of Kantian will and freedom,[22] and take up a far more modest use of *willing* or *wanting* to do something, closer to what Hume was talking about, and regularly expressed when we attribute an *intention* to someone.

Intentional behavior is understood as goal-directed: when we act intentionally, we intend to realize some goal, and that would seem the only proper use of "acting" and "action", the one Aristotle refers to when he says, at the very beginning of the *Nicomachean Ethics*:

> Every art and every inquiry, and similarly every action and choice, is thought to aim at some good; and for this reason the good has rightly been declared to be that at which all things aim.[23]

Insofar as my *unintentionally* scratching my head, or Thales' unintentionally (and, I would imagine, painfully) falling into a well, are not aiming at any goal (or even fail to reach whatever goal the agent had, such as watching the stars), they should not count as actions—hence the thread let loose a few pages ago can be tightened (for the moment; later it will be loosened again): an insane or ignorant person's behavior (or a machine's) does not qualify as acting.

Intentional, goal-directed behavior is guided by what, among Aristotle scholars, has come to be designated the *practical syllogism.* Though the Philosopher himself never uses that phrase, and what he says about its conjectured meaning is sketchy and spread around numerous unrelated works, there is enough there to draw some fairly assured deductions. Here is a passage from *Movement of Animals*:

> But how is it that thought is sometimes followed by action, sometimes not; sometimes by movement, sometimes not? What happens seems parallel to the case of thinking and inferring about the immovable objects. There the end is the truth seen (for, when one thinks the two propositions, one thinks and puts together the conclusion), but here the two propositions result in a conclusion which is an action—for example, whenever one thinks that every man

[22] But, before I do so, I note the connection between the Kantian notion of *Wille* and the conscience discussed in the last chapter. Expanding on the final note made there, as Kantian conscience monitors the rationality of my behavior, it also monitors the extent to which it could be called voluntary in a Kantian sense.

[23] Aristotle (1984), 1729 (1094a).

ought to walk, and that one is a man oneself, straightaway one walks; or that, in this case, no man should walk, one is a man: straightaway one remains at rest. And one so acts in the two cases provided that there is nothing to compel or to prevent. Again, I ought to create a good, a house is a good, straightaway he makes a house. I need a covering, a coat is a covering: I need a coat. What I need I ought to make, I need a coat: I make a coat. And the conclusion "I must make a coat" is an action. And the action goes back to a starting-point. If there is to be a coat, there must first be this, and if this then this—and straightaway he does this. Now that the action is the conclusion is clear.[24]

One from *On the Soul*:

Since the one premiss or judgement is universal and the other deals with the particular (for the first tells us that such and such a kind of man should do such and such a kind of act, and the second that this is an act of the kind meant, and I a person of the type intended), it is the latter opinion that really originates movement, not the universal; or rather it is both, but the one does so while it remains in a state more like rest, while the other partakes in movement.[25]

And one from the *Nicomachean Ethics*:

The one opinion is universal, the other is concerned with the particular facts, and we come to something within the sphere of perception; when a single opinion results from the two, the soul must in one type of case affirm the conclusion, while in the case of opinions concerned with production it must immediately act (e. g. if everything sweet ought to be tasted, and this is sweet, in the sense of being one of the particular sweet things, the man who can act and is not restrained must at the same time actually act accordingly).[26]

The following could be inferred from these passages:

1. (Intentional) action is the conclusion of a kind of reasoning analogous to the one labeled syllogism in theoretical contexts (the one letting us draw from the major premiss "All men are mortal" and the minor premiss "Socrates is a man" the conclusion "Socrates is mortal");
2. The major premiss of this "practical syllogism" contains some universal term, as in "All humans ought to take care of their health" or "I need some covering" (here "covering" is the universal), and the minor premiss applies the universal term to a particular, as in "For me to take care of my health requires that I do an annual physical" or "A coat is a covering."
3. The conclusion is an action, or at least commands an action: "I ought to do an annual physical" or "I need a coat."

What matters most for us: everything that counts as my acting at all, as opposed to just happening to me, orthogonally or even contrary to my plans— hence as opposed to something that could not called *mine*, besides not being an action—, occurs under the enlightened dictate of my (self-conscious) mind. I know general truths about myself and the world, and I know particular truths about my current situation, and from those truths I am (self-consciously) acquainted with there follows what I should

[24] Ibid., 1091–1092 (701a).

[25] Ibid., 690 (434)a.

[26] Ibid. 1812 (1147a).

be doing now—in fact, there follows what I *do* now. The Cartesian mind, therefore, has full control not only of the whole content of its experience but also of the whole purview of its actions. No wonder that analytic philosophers would have feasted on this. Here are a couple of distinguished representatives:

> "Practical reasoning" or "practical syllogism," which means the same thing, was one of Aristotle's best discoveries.[27]

> Aristotle distinguished between theoretical and practical syllogisms. His treatment of the latter is very scanty and unsystematic. The examples, of which he gives hints without elaborating them in detail, are a rather mixed bunch.[28]

I am not going to worry whether Aristotle could discover something he didn't even name, or whether his treatment of it is scanty and his examples a mixed bunch as a result of a lack of intellectual discipline (which von Wright and others are studiously going to counteract). For I am about to turn this topic into a different direction, to introduce which it is instructive to make one more remark about Aristotle.

The last passage from the *Nicomachean Ethics* quoted above is from Book VII, and is contained in an extensive and involved treatment of *akrasía*, most often translated today as "weakness of will" but more accurately rendered as "lack of control" (from the negative prefix *alpha* and the root of the verb "*kratéo*": to rule, to master)—a lack of that very control the mind is supposed by the Cartesians to keep over itself, its experience and now its actions as well. To give the punch line right away, it is contained within a discussion in which Aristotle shows how often and how persistently the practical syllogism *does not work*.

The acratic person knows, and approves of, the premisses of a practical syllogism, and is able to draw from them their logical conclusion, but then acts (often, though not always) in violation of that conclusion. A typical case is addictive behavior: I may know that I ought not to do things that damage my health, and that smoking damages my health, and I may correctly infer from those premisses that I ought not to smoke—only to then light up anyway. This condition was so puzzling for the Greeks that some of them (most notably, Socrates) went to the extreme of denying that it even existed:

> Now we may ask what kind of right belief is possessed by the man who behaves incontinently. That he should behave so when he has knowledge, some say is impossible; for it would be strange—so Socrates thought—if when knowledge was in a man something else could master it and drag it about like a slave. For *Socrates* was entirely opposed to the view in question, holding that there is no such thing as incontinence; no one, he said, acts against what he believes best—people act so only by reason of ignorance.[29]

But, pace Socrates, this sort of thing that he denied happens all the time, so much so that the rational, careful deliberator envisioned by supporters of the practical syllogism, old and new, seems to be largely a phantom of their imagination. Aristotle

---

[27] Anscombe (1957), 57–58.

[28] Von Wright (1963), 159.

[29] Aristotle (1984), 1809–1810 (1145b). Here "incontinence" and its cognates translate "*akrasía*" and its cognates.

has his own, interesting approach to the problem (it is not enough to reach a conclusion intellectually; one must be trained practically to see its relevance to her particular situations—so logical argument must defer here, according to "the father of logic," to example and exercise); but, leaving that aside, it seems that intentional behavior, if the practical syllogism is to account for it, gets reduced to a vanishing quantity. Considering how wide the range of addictive behavior is, and how resistant this behavior is to medical or psychological intervention, the self-conscious mind appears, as the presumed vehicle of goal-directedness, virtually impotent. Shall we then give up on goals, insofar as they demand an orderly arrangement of means in order to be attained, and think of human behavior as being "governed" by a chaotic array of conflicting passions, none lasting for long or providing any principle of organization?

Maybe not. For a more promising outlook I bring back the external observer who (I said earlier) is often in better touch with the contents of the self than the self is. In my last quote from Hume, he used this character to point out that the behavior of ours we judge free in the sense of indifferent often looks to "a spectator" as what can be inferred "from our motives and character." But Hume would think that behavior which is free in the sense of spontaneous is intentional in the ordinary (practical-syllogism) sense of the word: when I act spontaneously, I act in order to realize some of my intentions, and of course I know what they are. To get where I want to be, we need to generalize the spectator's role: an outside observer has *in general* a better understanding of what our intentions are than we do; even when *we* have such an understanding, it is because we adopt the role of an outside observer with respect to ourselves. Self-consciousness, here as elsewhere, has a derivative, not original—as well as partial, not complete—transparent and veridical access to the self. And, for that generalization, we can turn once more to Freud, this time to his *Psychopathology of Everyday Life*:

> At a certain point in the analysis of another woman patient I had to tell her that I suspected her of having been ashamed of her family during the period we were just then concerned with, and of having reproached her father with something we did not yet know about. She remembered nothing of the kind and moreover declared it was unlikely. However, she continued the conversation with some remarks about her family: "One thing must be granted them: they are certainly unusual people, they all possess *Geiz* [greed]—I meant to say '*Geist* [cleverness]'." And this was in fact the reproach which she had repressed from her memory.[30]

And here is Freud, in the same work, reporting on being an outside (and unsympathetic) observer of himself:

> An unpleasant part of my unconscious thoughts is disclosed by the following example. I may start by stating that in my capacity as a doctor I never consider my remuneration but only have the patient's interest in mind: that goes without saying. [Therefore, saying it makes is sound suspect.[31]] I was with a woman patient to whom I was giving medical attention in a period of convalescence after a serious illness. We had been through hard days and nights

[30] Freud (1953/1974), vol. VI, 64.

[31] And suspicion may be justified. Sándor Ferenczi (1988) writes: "I must remember some of Freud's remarks he let fall in my presence, obviously relying on my discretion: 'Patients are a rabble [...]. [...] Patients only serve to provide us with a livelihood and material to learn from. We certainly cannot help them" (93). Old-age cynicism? Possibly, but notice that *Psychopathology of Everyday*

> together. I was happy to find her improved; I painted a picture for her benefit of the delights of a stay in Abbazia [today's Opatija, a seaside resort town in Croatia, then part of the Austrian empire and of the so-called Austrian Riviera], and concluded by saying: "If, as I hope, you will *not* leave your bed soon…." This obviously owed its origin to an egoistic motive in the unconscious, namely that I should be able to continue treating this well-to-do patient some time longer—a wish that is entirely foreign to my waking consciousness [!?] and which I would indignantly repudiate.[32]

That the right perspective from which to ascertain a person's intentions in doing anything be that of an outside observer does not depend entirely on confirmation by psychoanalysis, on the other hand. If we turn to hermeneutics—the science of uncovering the meaning of one's pronouncements, texts, and actions (but, primarily, texts)—, a position that carries a lot of weight in it, and that would seem to give the most weight to the speaker's, writer's, or agent's self-consciousness, is the one making that meaning coincide with, or at least be determined by, her intentions. In the work done by the U. S. Supreme Court, which is to decide whether a law is consistent with the Constitution, this position is represented by *originalists*, who claim that what the Constitution means, hence what a law being discussed should be compared to, is what its framers intended.[33] But, since the framers are not around and we cannot question their self-consciousness about what they intended, we are left with making the sorts of conjectures any outside observer would have to make if she wanted to decide what someone else's intentions are.[34] Therefore, even when people are most deeply and professionally concerned with discovering (other) people's intentions, adopting an external perspective is often not only the primary, but also the *only* method for getting somewhere with it. (And I stigmatize again the simplemindedness of those analytic philosophers who, when discussing intentions, concentrate on oafish

---

Life was published in 1901, Ferenczi became associated with Freud as early as 1908, and he gives no indication in his diary of a change of mind on Freud's part on this matter. Most likely, I would think, greed was not (unconsciously) repressed but only (consciously) suppressed by Freud, hence was still consciously available to him; and his slip shows how little control (self-)consciousness had on his actions—and on the intentions expressed in them.

[32] Freud (1953/1974), vol. VI, 68.

[33] The starkest formulation of this position, in recent times, is in Bork (1971), where we read: "There appear to be two proper methods of deriving rights from the Constitution. The first is to take from the document rather specific values that text or history show the framers actually to have intended and which are capable of being translated into principled rules. We may call these specified rights. The second method derives rights from governmental processes established by the Constitution. These are secondary or derived individual rights" (17). Calamity follows if we veer from this course: "Unfortunately, the rhetoric of constitutional adjudication is increasingly a rhetoric about 'fundamental' rights that inhere in humans. That focus does more than lead the Court to construct new rights without adequate guidance from constitutional materials. It also distorts the scope and definition of rights that have claim to protection" (ibid.). Softer versions have been advocated by Antonin Scalia and Amy Coney Barrett.

[34] In Bork's language, we are left with inferring intentions from what "text or history show." Also, notice that, even if the framers *were* around, later (or even concurrent) statements by a party of what she intended by some action are often not taken to be definitive of her intentions or meaning, and are contrasted with what else her texts, her history, and in general her behavior show.

examples of the "The cat is on mat" variety, instead of looking at serious contexts where the intentional game is *really* played.)

But, if we sever the privileged relation self-consciousness was reckoned (by Cartesians of all stripes) to establish between a self and its intentions, and we find the locus of intentions in the guesswork spectators do on agents' behavior, the privileged role of *human* intentions and human goal-directedness goes away too. A kibitzer can speculate on the intentions of a poker or a bridge player, with far more brilliance and profundity than the player has available, as much as a political reporter can speculate on the intentions of Parliament during a combative session, or a voyager on a photographic safari on the intentions of a lion as it contemplates a herd of gazelles, or a biologist on the intentions of a group of trees as it salvages a stump. All of these people and speculations may be judged wrong, but only in the way in which a scientist and his theory are when they are found to be in conflict with the facts or when some other theory is found to give a better account of those facts, not in the way in which someone would be wrong in telling me what I think or want because the *cogito* is supposed to provide me with an infallible perception (different from his) of my own interior workings. And, though the traditional terminology may show strong resilience, and spectators' conjectures may still be portrayed as being about an agent's interior workings, mention of those workings is now vacuous: hypothesizing an intention or a goal is simply making predictions about how the behavior under scrutiny will evolve, and interior workings can drop out.[35] Whether the behavior's subject is a human, endowed with a traditional (immortal?) soul, or a zombie makes no difference. We might want to retain terms like "action" and "acting" for behavior that displays a certain amount of consistency (that a spectator is able to make some sense of); but it's a matter of degree, not of substance. None of the radical distinctions made or intimated by authors like Aristotle of Galiani hold water any more. Which also entails: a machine can have goals, can have intentions, can act.

There is one last wrinkle to consider: one last line of defense for the mentalists. Say that subject X goes through a self-conscious application of the practical syllogism, and acts accordingly. He draws the correct conclusion that he ought not to smoke and *doesn't* light up—feeling rather proud of thus exercising rational control on his behavior. Outside observer Y, however, connects this case of X's forbidding himself what would be for him an obvious pleasure with other similar renunciations on his part, and with incidents he knows about his growing up and experiencing stern parental discipline (especially with regard to weaning), and concludes that X's real intention in forfeiting cigarettes is not the rational one of promoting his health but the less-than-rational one of appeasing his sense of guilt. Shouldn't X's view of what he did prevail at least in this case? Shouldn't his first-person point of view, insofar as no inconsistency surfaces in it, trump Y's conjectural, third-person one?

The answer is No. Rationality—that is, the application of logical argument to the structuring of conduct—is a tool and also a weapon: since no argument can

---

[35] On p. 40 I said that what goes through others' minds never enters the actual proceedings of figuring out what they mean "however it might be vacuously mentioned." The current discussion should make that reference clearer.

certify the truth of its premisses, it can be used as much to show the fruitfulness of good premisses as to dodge the spuriousness of bad ones. The *appearance* of consistency urged by *rationalization* is indeed, Freud warns, a major obstacle to deciphering behavior. He brings this up with special emphasis in regard to what he calls *secondary revision* in a dream, that arranges the different components of the dream content, distorted as they are by the dream-work, into as much as possible of a coherent story. In the following passage from *The Interpretation of Dreams* he traces an intriguing parallel between this operation and everyday waking thought, and points out the risk involved in both of getting the wrong picture of things:

> [It is] highly probable that the psychical function which carries out what we have described as the secondary revision of the content of dreams is to be identified with the activity of our waking thought. Our waking (preconscious) thinking behaves towards any perceptual material with which it meets in just the same way in which the function we are considering behaves towards the content of dreams. It is the nature of our waking thought to establish order in material of that kind, to set up relations in it and to make it conform to our expectations of an intelligible whole. In fact, we go too far in that direction. An adept in sleight of hand can trick us by relying upon this intellectual habit of ours. In our efforts at making an intelligible pattern of the sense-impressions that are offered to us, we often fall into the strangest errors or even falsify the truth about the material before us.[36]

In this other passage, Nietzsche applies the same sobering diagnosis to himself (relative to his waking life):

> Perhaps [...] I might be reproached with having employed a certain amount of "art," a certain amount of false-coinage: for example, that I knowingly-willfully closed my eyes before Schopenhauer's blind will to morality at a time when I was already sufficiently clear-sighted about morality; likewise that I deceived myself over Richard Wagner's incurable romanticism, as though it were a beginning and not an end; likewise over the Greeks, likewise over the Germans and their future—and perhaps a whole long list could be made of such likewises?—Supposing, however, that all this were true and that I was reproached with it with good reason, what do *you* know, what *could* you know, of how much cunning in self-preservation, how much reason and higher safeguarding, is contained in such self-deception—or of how much falsity I shall *require* if I am to continue to permit myself the luxury of *my* truthfulness? ... Enough, I am still living; and life is, after all, not a product of morality: it *wants* deception, it *lives* on deception.[37]

I have heard some jackasses claim that most of our ordinary beliefs must be true because otherwise the inexorable laws of evolution (mutation and selection) would have determined our extinction. To them I would respond with Campanella's words:

> [O]ur knowledge of man, of wine, and of our things is sufficient for a generic wisdom, needed for the conservation of life, as for a worm it is sufficient to know the small parts of cheese where it was born and which it eats, but it is not sufficient to simply know what is wine, bread, man and their origin, and the elements, and the world and the one who governs it. In the same way the worm does not know everything about the cheese, nor its essence, nor its place in the world nor the elements or the causes whence it comes.[38]

---

[36] Freud (1953/1974), vol. V, 499.

[37] Nietzsche (1996), 5–6.

[38] Campanella (1994), 67; translation mine.

As long as we know enough about the cheese where we live to move efficiently in it, no other "truth" is required for our survival, and for our keeping an edge within the struggles of natural selection. So we should look askance at our presumptuous efforts to put the world of our experience and action into some rational order. They might have no better value than the similar efforts by which we try to bring order into the pandemonium that condensation, displacement, and symbolism have made of the root meaning of a dream. Nietzsche's passage adds a further epicycle to this hellish configuration: there may be a rationality to what is going on, but uncovering it may require *rejecting* any credibility our own (self-conscious) rationalizations aspire to. In the context of that passage, he acknowledges that, in making his most outrageous statements, he has always been looking for company:

> What I again and again needed most for my cure and self-restoration [...] was the belief that I was *not* thus isolated, not alone in *seeing* as I did—an enchanted surmising of relatedness and identity in eye and desires, a reposing in a trust of friendship, a blindness in concert with another without suspicion or question-marks, a pleasure in foregrounds, surfaces, things close and closest, in everything possessing colour, skin, and apparitionality.[39]

So he deluded himself (the *reality* of self-deception is as endemic for me as the *paradox* of it is endemic for the Cartesians) that he had something in common with Schopenhauer, or Wagner, or the Greeks; but what if his own holding on to life depended on that delusion, on that fake sense of commonality? Then it would have been rational for him to adopt an irrational posture: to convince himself of what was eventually to be proven false—of what he, other than for his rationalizations (for his employing a certain amount of "art"), should have *known* was false.

Here, finally, is where I stand: It might be that, at times, faithful adherence to a practical syllogism by a self-conscious agent is the best explanation of what she wills—of what her intentions are. And it might be that, at such times, outside observers' interpretations of her behavior—insofar as they don't fit with the agent's—must be rejected. But none of this ever leaves the realm of conjecture, of hypothesis, of guesswork; self-consciousness does not grant the agent any direct possession of the truth. The agent is going to be wrong as often as she is right. Therefore, when dealing with agents (including other animals, plants, and machines) to which we despotically deny self-consciousness, nothing much changes: within the same conjectural realm, and with the same level of (un)certainty, we can attribute intentions to them, we can guess what their goals are, what they want or will to do. The only difference is that we don't have to deal with a vociferous, vainglorious bother who unbendingly protests that, *of course*, he knows perfectly well what he wants and what he does.

[39] Nietzsche (1996), 5.

References

Anscombe, Gertrude Elizabeth Margaret. *Intention*. Oxford: Blackwell, 1957.

Aristotle. *The Complete Works of Aristotle*, edited by Jonathan Barnes. Princeton NJ: Princeton University Press, 1984.

Bencivenga, Ermanno. *Ethics Vindicated: Kant's Transcendental Legitimation of Moral Discourse*. New York: Oxford University Press, 2007.

Bork, Robert Heron. "Neutral Principles and Some First Amendment Problems." *Indiana Law Journal* 47 (1971), 1–35.

Campanella, Tommaso. *Metafisica*, vol. I, edited by Paolo Ponzio. Bari: Levante, 1994.

Dennett, Daniel. *Elbow Room: The Varieties of Free Will Worth Wanting*. Cambridge MA: The M.I.T. Press, 1984.

Ferenczi, Sándor. *Clinical Diary*, translated by Michael Balint and Nicola Zarday Jackson. Cambridge MA: Harvard University Press, 1988.

Freud, Sigmund. *The Standard Edition of the Complete Psychological Works of Sigmund Freud*, in twenty-four volumes, edited by James Strachey. London: The Hogarth Press, 1953/1974.

Galiani, Ferdinando. *Opere*, edited by Furio Diaz and Luciano Guerci. Milano: Ricciardi, 1975.

Hume, David. *A Treatise of Human Nature*, edited by Lewis Amherst Selby-Bigge, second edition revised by Peter Harold Nidditch. Oxford: Clarendon Press, 1978.

Kant, Immanuel. *Anthropology from a Pragmatic Point of View*, translated by Victor Lyle Dowdell. Carbondale IL: Southern Illinois University Press, 1978.

Kant, Immanuel. *Practical Philosophy*, translated by Mary Jeanne Gregor. Cambridge UK: Cambridge University Press, 1999.

La Mettrie, Julien Offray de. *Man as Machine*, translated by Gertrude Carman Bussey and Mary Whiton Calkins. Chicago: Open Court, 1912.

Nietzsche, Friedrich. *Human, All Too Human: A Book for Free Spirits*, translated by Reginald John Hollingdale. Cambridge UK: Cambridge University Press, 1996.

Nietzsche, Friedrich. *Beyond Good and Evil: Prelude to a Philosophy of the Future*, translated by Judith Norman. Cambridge UK: Cambridge University Press, 2002.

Sapolski, Robert Morris. *Determined: A Science of Life without Free Will*. London: Penguin, 2023.

von Wright, Georg Henrik. *Norm and Action: A Logical Enquiry*. New York: Humanities Press, 1963.

Chapter 8
Meaning

Abstract Traditionally, meaning was closely connected with intentions: it was what one meant, or intended, by an utterance. And what was meant was taken to be objects or facts in the outside world, or private experiences accessible to the *cogito*. But de Saussure introduced an alternative view, according to which language operates by the differences it sets among its various elements; "in language there are only differences." Formalists like Hilbert go in the same direction when they propose that the basic (primitive) terms of a theory cannot be defined (as they were in Euclid's geometry), not explicitly at least: their meaning is, rather, implicitly determined by the relations the theory's axioms establish among them. This view too can be generalized to all use of language: whether spoken or written, someone's idiolect acquires meaning, and can be understood, only by inquiring into how the speaking or writing of it establishes relations among its elements. Viewing language differentially makes one appreciate the similarity between its use by humans and by AI. For the latter works by absorbing huge amounts of text and uncovering relations within them. It is literally true for it, as Derrida would say, that there is no outside to the text; but then the same is true for humans as well, if by "text" we understand not just something written but the whole context of one's behavior. As far as meaning goes, our approach is the same whether we deal with machines or with humans.

Keywords Meaning · Difference · Formalism · Implicit definitions · Text

It's worth our while to insert a footnote here, or if you will a digression, connecting the theme of intentions with that of language, treated in Chapter 4. Here we find de Saussure parting ways from Aristotle, and from most of the tradition reaching to himself, in ways that are of special relevance to the main topic of this book.

According to that tradition, the *meaning* of a linguistic utterance is what the speaker *means*—that is, *intends*—to convey. The meaning of "The cat is on the mat" is the fact (real or unreal) that the cat is located there (or, if you are Frege, a truth-value), which the speaker intends to communicate to his hearers; the meaning of an article of the U. S. Constitution (if you are an originalist) is what the Founding Fathers intended by writing it. If you go the way traveled in the previous chapter, an outside

E. Bencivenga, *The Philosophy of (Artificial) Intelligence*,
https://doi.org/10.1007/978-3-032-09347-9_8

observer would in general be better equipped to determine what that intention, hence that meaning, is; and she would be no less authorized to such a determination with a machine than with a human.

But we can get even more sympathetic to machines, especially to those kinds (computers and computer networks) that are closest to our concerns, while also getting clearer about the technique to be used in fathoming the intentions (and meanings) of whatever we might be observing as spectators. For consider this passage from de Saussure (1959):

> Everything that has been said up to this point boils down to this: in language there are only differences. Even more important: a difference generally implies positive terms between which the difference is set up; but in language there are only differences *without positive terms*. (120)

The point can perhaps best be seen within phonetics. If you stick to Aristotelian definitions by genus and differentia, you will conceive of a meaning as constituted by an accumulation of traits. "Human," for example, will mean *rational animal*—or, in more effusive terms, *rational self-moving animate corporeal substance*. But suppose now that I want to define the phoneme "*t*." No accumulation of traits will do the job, since your pronunciation of *t*s may have little in common with mine, and a lot in common with the way I pronounce *d*s. Which may give me a little trouble as we begin to interact, until I penetrate your system, recognize the relevant *differences*, and can then successfully deal with you. What matters for successful linguistic exchanges is not any number of positive features the sounds of individual phonemes must have, but the ways they *differ* from one another—the subtle or not-so-subtle contrasts that exist among them. In the passage above, de Saussure is generalizing this example to the totality of (any) language: "in language there are only differences"; and, what counts even more, not differences from some positive basis which is given to begin with. The contrasts among the elements of (a) language are *all there is to (a) language*.

Much the same point was made around the same time, in the philosophy of mathematics, by formalists like David Hilbert. Euclid (1956) opens with Aristotelian definitions of the main concepts to be employed. Among them we find:

1. A **point** is that which has no part. (155)
2. A **line** is breadthless length. (158)
3. A **surface** is that which has length and breadth only. (169)

But it's not clear that such definitions are adequate. The Philosopher says, in the *Topics*, that a first sense in which a definition is incorrect is when it uses "obscure language (for the language of a definition ought to be the very clearest possible, seeing that the purpose of rendering it is to make something known)"[1]; and do we know what "breadthless length" is any better than we know what a line is? Does Euclid's "definition," then, make a line known to us?

To overcome this difficulty, Hilbert's 1899 axiomatization of geometry posits that some terms just *cannot* be defined—not, at least, as Aristotle would: they cannot

---

[1] Aristotle (1984), 235 (139b).

be given *explicit* definitions; they are *primitive*.[2] What happens, rather, is that they are *implicitly* defined by the relations the axioms establish between them and other primitive terms. Points, straight lines, and planes are *any* sorts of entities *whatever* that "differ" from one another in the ways stated by the axioms. As long as they bear those relations, entities don't have to *resemble* what we are used to calling "points," "straight lines," and "planes" in order to qualify for such roles.[3] In a quote often attributed to him, "One must be able to say, instead of 'points, straight lines, planes,' 'tables, chairs, beer mugs'": if tables, chairs, and beer mugs were related to each other as the theory assumes points, straight lines, and planes to be related, then the theory, in speaking of points, straight lines, and planes, would also be speaking of tables, chairs, and beer mugs—no matter what the "intentions" were of the person who devised it.[4]

Hilbert's view, much like de Saussure's, can be generalized to the whole use of language. In trying to understand a text, dictionary definitions of key words can at best provide some initial orientation; then we must become familiar with how those words, or indeed any words, are "axiomatized" by statements in the text, which implicitly define them. In fact, the mistake most often, and most ruinously, done by interpreters of texts (including philosophical ones) is that of importing into them "meanings" established outside of them. Therefore, Jacques Derrida (1976) is right when saying "*Il n'y a pas de hors-texte*" ("there is nothing outside of the text," "there is no outside-text," 158)—though he is not right in seeing this as evidence of a priority of written language over the spoken variety. Exactly the same "interpretation from within" is in order with an oral conversation as is with something we read: if we want to avoid severe misconstructions of what someone is saying, and possibly serious quarrels, we'd better figure out what any of her words mean from the context of all her other utterances, past and present.

This understanding of meaning is more congruent with seeing language as room for play than the Aristotelian one. If the meaning of "straight line" or "beer mug" is a specific class of objects out there—a "transcendental signified," as Derrida would call it,[5] independent of language—then the play of language is limited in its use of these phrases by the requirement that the connection with those specific objects be maintained. But, if there is no outside to language, then the play is freer: stories can be told where straight lines are (also?) circumferences (that decided to close themselves in an embrace?) and beer mugs are souls of deceased drunkards. (Notice that we can tell such stories anyway, which shows how wrong the Aristotelian view

---

[2] See Hilbert (1903).

[3] In formal-logical terms, most interesting theories are non-categorical, that is: they have a number (even a very large number) of non-isomorphic models. To give an idea of how large this number can be, Ehrenfeucht (1958) proved that elementary arithmetic has $2^{\aleph_0}$ (as many as the power of the continuum) non-isomorphic models of any infinite cardinality.

[4] The sentence here attributed to Hilbert is not to be found in any work of his, but is reported by Otto Blumenthal as said by him during a conversation in the *Lebensgeschichte* (biography) of Hilbert that Blumenthal appended to Hilbert (1935), 402–403.

[5] See ibid., 49. Notice that Derrida's use of the term "transcendental" is different from mine (or, for that matter, Kant's).

of language is.) And, if I want to catch how *you* play with (and in) language, I need to disregard anything outside and wiggle my way into that play of yours, learn to resonate with what moves are made there; then (who knows?) I might even be able to extend them and refine them.

This notion of wiggling one's way into someone else's play also gives us a more precise picture, as I pledged, of how specifically we inquire into what the Founding Fathers meant by a sentence of the Constitution, or what a friend meant by a remark that at first left us puzzled. What we do is compare that sentence, or that remark, with "text or history": what history tells us about the Founding Fathers, what my memory tells me about my friend's whole behavior, what other sentences, or other remarks, were offered by these agents. All of this in order to crack their idiolect (their particular form of play), to set up the proper relations of difference among their terms, to unveil the axioms that express those relations—and of course it's all conjectural anyway, but it's the only thing we can, and will, do.

Now look at AI. For it, there literally is nothing outside of the text: what it works with is massive datasets of text, within which it uncovers relations and from which it proceeds to construct significant sequences of its own text. Searle would say that, as it does so, it does not *understand* the *meaning* of what it's doing (or "saying"), and I have been exposing the delusional character of that claim; but the "differential" view of language offers another daunting weapon in this confrontation. For, if we accept that view, neither do you, in getting what I "mean" by the use of "beer mug," capture it by anchoring it solidly in non-linguistic beer mugs, and nor do even I. What I "mean" is determined by how "beer mug" relates to all other words and phrases in my idiolect, and even I could be surprised if I ran a search of those relations. I am no more an owner of my meaning (of my play) than AI is; I cannot get assured of it by pointing to an object outside or to some experience inside of me; it is all at the same time outside (of me) and inside (of language)—as much for me as for any language-using machine.

References

Aristotle. *The Complete Works of Aristotle*, edited by Jonathan Barnes. Princeton NJ: Princeton University Press, 1984.

Derrida, Jacques. *Of Grammatology*, translated by Gayatri Chakravorty Spivak. Baltimore MD: The Johns Hopkins University Press, 1976.

de Saussure, Ferdinand. *Course in General Linguistics*, translated by Wade Baskin. New York: Philosophical Library, 1959.

Ehrenfeucht, Andrzej. "Theories Having at Least Continuum Many Non-Isomorphic Models in Each Infinite Power" (abstract). *Notices of the American Mathematical Society* 5 (1958), 680.

Euclid. *The Thirteen Books of Euclid's Elements*, edited by Thomas Little Heath. Mineola NY: Dover, 1956.

Hilbert, David. *Grundlagen der Geometrie*, revised edition. Leipzig: Teubner, 1903.

Hilbert, David. *Gesammelte Abhandlungen*, vol. III. Berlin: Springer, 1935.

Chapter 9
Scale and Shape

Abstract Throughout this book, I have been fighting anthropocentrism and anthropomorphism, especially insofar as they impede our recognition of intelligence, soul, language, thought, consciousness, and will in beings that do not look like us. Here I focus on two specific aspects of this problem: we have a hard time giving credit to beings that are of a different size from us, and that have a shape different from our own. Size difference works (or, rather, doesn't) in both directions. When dealing with beings vastly larger than ourselves, like States, companies, or the "military-industrial complex" denounced by U. S. President Eisenhower in his farewell address to the nation, we tend to think of them in terms of their human members, and not to see the logic followed by the whole complex, which most often transcends, and even counters, that of any human or group of humans. Same thing when we deal with beings much smaller than ourselves, like intelligent proteins: we tend to think of their intelligence as a figure of speech, instead of an obvious reality. A State, a company, or the military-industrial complex are also a different shape from a human body—which for Aristotle was a perfect reflection of the structure of the cosmos. Therefore, if we even recognize their intelligence and other "spiritual" qualities, we tend to think of them as resulting from arrangements similar to our own: a central unit (a brain) receiving information and dispensing instructions to all the rest. But that is just lack of imagination: an army, a forest, a company, a city, or an octopus have a diffuse intelligence—they play, learn, speak, think, and will with their entire being, in ways that were brought to general attention by the work of Deleuze and Guattari on rhizomes. And so do computer networks.

Keywords Anthropocentrism · Anthropomorphism · Size · Shape · Rhizome

On January 17, 1961, three days before the inauguration of his successor John Fitzgerald Kennedy and after half a century of public service, President Dwight David

E. Bencivenga, *The Philosophy of (Artificial) Intelligence*,
https://doi.org/10.1007/978-3-032-09347-9_9

Eisenhower delivered a joint radio and television farewell address to the American people.[1] It contained the following ominous warning:

> [W]e have been compelled to create a permanent armaments industry of vast proportions. Added to this, three and a half million men and women are directly engaged in the defense establishment. We annually spend on military security more than the net income of all United States Corporations. This conjunction of an immense military establishment and a large arms industry is new in the American experience. The total influence—economic, political, even spiritual—is felt in every city, every State house, every office of the Federal government. We recognize the imperative need for this development. Yet we must not fail to comprehend its grave implications. Our toil, resources and livelihood are all involved; so is the very structure of our society. In the councils of government, we must guard against the acquisition of unwarranted influence, whether sought or unsought, by the military-industrial complex. The potential for the disastrous rise of misplaced power exists and will persist. We must never let the weight of this combination endanger our liberties or democratic processes. We should take nothing for granted. Only an alert and knowledgeable citizenry can compel the proper meshing of the huge industrial and military machinery of defense with our peaceful methods and goals, so that security and liberty may prosper together. (13–16)

This is a statement of enormous historical, political, and—last but not least—transcendental[2] significance. First, coming from a president who had been one of the highest-ranked generals ever[3] and who had understood the value of a serviceable (and highly expensive) highway system for military purposes and, as a president, had realized it,[4] it cautions the public to the fact that military spending had gone through a dramatic increase and could get out of hand. This was to be proven true: U. S. military spending was $47 billion in 1960, less than half a trillion dollars in today's money, and is now twice that amount, and over half of total federal discretionary spending (besides being over one third of all military spending in the world). Second, with this deluge of money comes tremendous power, which may ruin the balance among distinct branches of government that is the most reliable bastion of freedom (and the citizens are exhorted to understand the problem and act on it—something they have hardly been doing). Third, and crucially for us, the statement does not

---

[1] A reading copy of the whole speech, from which I quote below, can be found at the website of the Dwight. D. Eisenhower Presidential Library, at https://www.eisenhowerlibrary.gov/research/online-documents/farewell-address.

[2] In this regard, it is relevant to point out the reference to a "spiritual" influence.

[3] Eisenhower was a five-star general, a rank introduced in 1944 and retired in 1981, after the death of Omar Nelson Bradley. Only four other people (including Bradley) had that rank. A handful of people in history were granted the equivalent rank of general of the army, and three more (two of them posthumously: George Washington and Ulysses Grant) the even higher rank (equivalent to six-star) of general of the armies.

[4] Eisenhower was deeply impressed, as a young officer, by the sorry state of U. S. roads during a trip he had taken across the country in 1919 to ascertain the difficulties of transporting military equipment (it took them sixty-two days to cover 3200 miles) and, on the contrary, by the efficiency of the German *Reichsautobahn* during World War Two. (Military history would have also taught him how important the Roman roads were for the rapid movement of the legions.) In recognition of his role in making it happen in the U. S., the Interstate System there bears his name and so does the tunnel that might be regarded as its capstone, completed in Colorado in 1973 at an altitude of over eleven thousand feet.

refer to the work done, or the risk posed, in this regard by individual humans: its target is a larger entity—“the military-industrial complex,” “the huge industrial and military machinery of defense”—that may acquire “unwarranted influence” and, with its “weight,” “endanger our liberties or democratic processes.” A particularly suggestive phrase is “whether sought or unsought”: this gargantuan, dangerous entity might not *seek*, in any way that we would recognize as such, to acquire unwarranted influence—it is its very existence, and its proceeding under its own steam, that can come to that.[5]

The two main attitudes I have been fighting in this book, not accidentally, have the same Greek root “*ánthropos*” (“man”), and it’s time to devote some concentrated attention to them. There is *anthropocentrism*: the view that humans are the center of the universe—at the same time its noblest member and its ultimate goal. In Biblical terms, humans were created in a god’s image and the rest of creation was assigned the task of serving them (while they were assigned the task of ruling over it). Current slogans like “Drill, baby, drill!” are evidence that, despite Copernicus’, Darwin’s, Freud’s, and countless other blows to our narcissism, we haven’t strayed too far from this ancient, tired tale (which, more than anything else it’s doing, is *self*-serving). Related to its childish pretense is *anthropomorphism*: according to it, anything we could meaningfully relate to, or engage with, either as a companion or an adversary, must look like us, must be similar to a human. In the old days, this view extended to natural phenomena, which were taken to be the result of action by (again!) a variety of gods—that is, essentially, by attractive, enviable projections of ourselves, as current superheroes are. Zeus was responsible for lightning, Poseidon “the earth-shaker” for earthquakes, and Demeter for the growing of crops. Today those gods are no longer in business; but that is because we no longer see Nature as unitary and integrated (if only as a battlefield for the confrontation of divine agents). Nature is just a bunch of things going on, following laws that we do our best to guess, but is not itself going *anywhere*: is not being for us a partner in a conversation, an ally or an enemy in a joint enterprise or a principled conflict, the carrier of plans that might favor or hamper our own.

In the social and political realm, we do acknowledge the existence of goals and projects—by States, corporations, or associations—which, at any one time, might find themselves in opposition to the goals or projects of individual humans; but then anthropomorphism raises its ugly head and issues in the banality of conspiracy theories. It’s still humans doing it: a group of them, perfidious and secretive, are unfurling a scheme at the expense of us all in order to exponentially increase their income and domination. Which is laughable: after all I said about the deceptiveness of self-ascribed intentions, I will admit that there may be people devising treacherous stratagems in their petty minds, but I would invite them (and the ones watching them)

[5] Notice also other impersonal ways of referring to this entity: “[t]his conjunction,” “this development,” “this combination.”

to reflect on the wisdom of Jean de La Fontaine's 1668 fable "The Coach and the Fly."[6]

> Up a long dusty hill, deep sunk in sand,
> Six sturdy horses drew a Coach. The band
> Of passengers were pushing hard behind:
> Women, old men, and monks, all of one mind.
> Weary and spent they were, and faint with heat;
> Straight on their heads the sunbeams fiercely beat.
> In the hot air, just then, came buzzing by,
> Thinking to rouse the team, a paltry Fly.
> Stings one, and then another; views the scene:
> Believing that this ponderous machine
> Is by his efforts moved, the pole bestrides;
> And now upon the coachman's nose he rides.
> Soon as the wheels begin again to grind
> The upward road, and folks to push behind,
> He claims the glory; bustles here and there,
> Fussy and fast, with all the toil and care
> With which a general hurries up his men,
> To charge the broken enemy again,
> And victory secure. The Fly, perplexed
> With all the work, confessed that she was vexed
> No one was helping, in that time of need.

To be sure, some are sitting pretty today right at the corner where world-historical forces are unleashed, and as a result are making themselves unprecedentingly rich and admired; but it's not their doing. The coach is being driven by its six sturdy horses, and will go its way no matter what; they are the flies buzzing around and giving themselves an air of importance—as La Fontaine points out at the end, in giving his story's moral:

> So, certain people give themselves great airs,
> And meddlers mix themselves with one's affairs,
> Try to be useful, worry more and more,
> Until, at last, you show the fools the door.

It is, to a major extent, a question of scale. Humans are midsize objects (or, at least, they look midsize to themselves), and they are comfortable identifying, and relating to, other objects of their approximate size. They conduct their everyday lives in a world of tables and chairs, automobiles and trucks, color TVs and cellular

[6] See it at https://www.literature.com/book/the_coach_and_the_fly_2598. As with many of his fables, La Fontaine was inspired here by ancient sources, specifically by Phaedrus' "The Fly and the Mule," (in Phaedrus (1845), 87–88).

phones; when they get in front of vast spectacles (a sunset on the ocean, a mountain range, a rock concert) which are to be appreciated as a whole, they are overwhelmed by emotion and get into a kind of ecstasy—from the Greek "*ékstasis*," which means "displacement," "entrancement," "rapture": they literally get out of themselves (even if the concert is rubbish: it's not the fare that does it; it's the scale).[7] In the eighteenth and nineteenth century, a bulky literature (that included, among many others, Edmund Burke and Schopenhauer) blossomed discussing the notion of *sublime* experiences: majestic and scary ones that provided pain and pleasure at the same time.[8] When Kant deals with the matter in the *Critique of the Power of Judgment*,[9] he makes the sublime into a humbling of our empirical nature, a demonstration of its utter inadequacy—which, however (here is where the duality of pain and pleasure comes from), also brings to light the supreme dignity of reason: according to him, we (our empirical nature) are (is) being humiliated by ourselves, by our own *rational* nature.

This discomfort with scales different from (approximately) our own makes it hard to figure out, in a large number of circumstances, what the subject is we should be focusing on. I have already talked about Mancuso's views: a tree is (more or less—though rather more than less) our size, hence we are inclined to see it as a subject of relevance in the field of botanical studies; but what if we learn nothing that way, what if, by going that way, it remains distressingly mysterious how a tree that is supposed to be stone-dead is nonetheless alive, and the right point of view from which everything becomes clear is the one that sees the forest—a *much* bigger entity than we are—as the subject at work there?

A similar problem arises in evolutionary biology, from the very beginning. Here is Darwin, in *On the Origin of Species*:

> The great difficulty lies in the working ants differing widely from both the males and the fertile females in structure, as in the shape of the thorax and in being destitute of wings and sometimes of eyes, and in instinct. As far as instinct alone is concerned, the prodigious difference in this respect between the workers and the perfect females, would have been far better exemplified by the hive-bee. If a working ant or other neuter insect had been an animal in the ordinary state, I should have unhesitatingly assumed that all its characters had been slowly acquired through natural selection; namely, by an individual having been born with some slight profitable modification of structure, this being inherited by the offspring, which again varied and were again selected, and so onwards. But with the working ant we have an insect differing greatly from its parents, yet absolutely sterile, so that it could never have transmitted successively acquired modifications of structure or instinct to its progeny. It may well be asked how is it possible to reconcile this case with the theory of natural selection?[10]

[7] These days, they would try to manage their anxiety by reducing the experience to a scale they can control: by taking pictures of the spectacle, probably on their phones. The pictures, being midsize objects, cannot of course do justice to the experience, but that's not what matters to them: they need to get back in control of themselves, dealing with something at their scale.

[8] In the Italy of the Grand Tours the "*orridi*" (horrific gorges and precipices) became very popular among travelers. Famous is the Orrido of Bellano, mentioned by (among others) Stendhal (1973), 138.

[9] Kant (2002), 128ff.

[10] Darwin (1996), 192–193.

Darwin himself saw a way out of the "difficulty," which, he said, would disappear if we "remembered that selection may be applied to the family, as well as to the individual."[11] In other words, there would not be a problem if in this case, and many others, the subject of evolution were taken to be not the single ant but the colony (not the tree, but the forest). Colonies survive or do not; they mutate and transmit their mutations to descendant colonies; they will be favored in terms of survival (favored by selection) if their structure (including their social structure) is more adaptive to the environment—and if that involves that a large number of their members be sterile, so be it. But this simple idea had to struggle with the "modern synthesis" of evolution theory and Mendelian genetics that has dominated the field since the mid-twentieth century.[12] If individuals are, through their genes, the carriers of evolution, we are back to square one: when a sterile working ant dies, leaving no progeny, its genes go with it, and so does whatever advantage its work supplied to the colony.

More than a century after Darwin's masterpiece, Hamilton (1964) proposed the theory (which became dominant) of inclusive fitness, whereby an individual will favor not only its own progeny but also, to some extent, progenies it is related to (not only, say, one's own children but also one's nephews and nieces)—and that has been touted as offering a genetic basis for the existence of altruism. In terms of our problem, it is still individuals doing it, but they have a more expansive, enlightened notion of their personal interest. Give it another four decades, and two researchers at McGill University "discovered molecular signals [coming from the social environment] that can maintain social harmony in ants by putting constraints on their fertility." "[H]igh levels of reproductive constraint has most likely evolved as a consequence of selection at the colony level to reduce or eliminate any potential conflict over worker reproduction, therefore maintaining harmony and colony efficiency in advanced ant societies" (Khila and Abouheif (2008)). "Advanced," yes, because not all colonies are winners here: "'More primitive, or ancestral, ants tend to have smaller colony sizes and have much higher levels of conflict over reproduction than the more advanced species,' Abouheif explained. 'That's because the workers have a much higher reproductive capacity and there is conflict with the queen to produce offspring.'" "More than 150 years ago, Charles Darwin had an idea and now he's been proven right." "This study supports Darwin's prescient ideas ["that in the case of ant societies natural selection applies not only to the individual, because the individual would never benefit by cutting its own reproduction, but also to the family or group"], and provides a molecular measure of how an entire colony can be viewed as a single or 'superorganism.'"[13] It took a while; and it did, I surmise, because humans have

---

[11] Ibid., 193.

[12] The term "modern synthesis" was introduced by Huxley (1942). Versions of the theory were offered by Dobzhansky (1937), Mayr (1942), and Stebbins (1950).

[13] The unattributed quotes are from a communication reporting the discovery, issued by McGill Newsroom on November 19, 2008, and entitled "Worker Ants of the World, Unite! You Have Nothing to Lose but Your Fertility." See it at https://www.mcgill.ca/newsroom/channels/news/worker-ants-world-unite-you-have-nothing-lose-your-fertility-102785. See also Abouheif's (2022) comprehensive account of his ant work. (https://www.abouheiflab.org announces that the Abouheif

a hard time seeing something at the scale of an ant colony (or a forest) as a viable subject. So they have a hard time finding what they are *not* looking for.

The military-industrial complex is far more massive than an ant colony or a forest, hence it will be so much harder to see it as a subject, and so much more tempting to attribute its behavior to the single ants that sit in its offices, drive its trucks, or fire its missiles—or to some cluster of them joined by common interests or ambitions. It is going to be passed over in silence that the individual ants come and go, that they often have personal goals contrary and even detrimental to the complex, that they often make serious blunders when they take themselves to be working for the benefit of the complex, and yet the complex continues to make progress and is not annoyed by those faux pas more than an ant colony would be if a worker were devoured by a sparrow, or a group of them were sidetracked by a sudden rainstorm. Oblivious to all these recurring, evident facts, humans look at the pertinent contexts and what they *see* there is men and women much like themselves moving about or standing still, wheeling and dealing, and maybe office desks and trucks and missiles; they don't see *a complex*, or more appropriately a small part of a complex working *through* those men and women and desks and trucks and missiles. It requires significant imagination to see things that way; and, as Bruno de Finetti griped once, "Our lack of imagination is truly appalling."[14]

Natural evolution (as it works among biological species as well as among companies and other social entities) is a grandiose application of play, in the sense of play I clarified at the beginning: it is a venturesome trying out of steps away from the beaten paths, which may bring about disaster (monstrosities) but may also land a subject in much more commodious quarters. And the issue of scale I have been exploring here reminds us of something else I also said at the beginning, and will become crucial in the next chapter: play may be happening simultaneously at various levels (at different scales), and it may be instructive and helpful at one level while it is of no value, or even counterproductive, at others. The little games the little men and women members of the complex play with their peers or with themselves may end up destroying their careers or their health, *and yet* the complex may well learn from them—and from their destroyed careers and health—precious lessons that advance its own prosperity.

Difference of scale does not only obstruct vision when things are too large; it does at the other end, too—when they are too small. It's hard to see, for example, that something as tiny as a protein can manifest intelligence. But it does, and, as witness to that, I will quote now from the recent article "'Intelligent' Proteins"[15]:

> The concept of "protein intelligence" can be seen as an evolutionary advancement in understanding protein behavior from the traditional mechanistic view, which treats proteins as static, pre-programmed molecular machines. Instead, it posits that proteins exhibit a form of

lab is moving to China's Zhejiang University on August 1, 2025, and is recruiting graduate students and postdocs.)

[14] De Finetti (2006), 74; translation mine.

[15] Tripathi et al. (2025).

minimal intelligence, which is characterized by their ability to integrate information, adapt to environmental changes, and exhibit memory-like properties. (3–4)

Protein intelligence, as discussed in this article, suggests that proteins possess the ability to adapt and learn from a changing environment, memorize, anticipate future changes, solve problems, identify patterns, recognize patterns, actively shape or modify their environment when necessary, and control and modulate the behavior of their interaction partners. (22)

The authors state that all the features they have identified are "clearly anthropomorphic" (ibid.), in the sense that they are the sorts of features we normally attribute (only) to humans; but of course proteins do not *look* like humans, which may be the reason why "these actions were generally not explicitly linked to protein intelligence, apart from a largely metaphorical usage of the term" (23). They hope that "the arguments presented in this article offer a new perspective on proteins and their activities" (ibid.), and so do I; but, for that to happen, it is important to have a conception of intelligence which is ample enough to capture more than humans in a strict, not just "metaphorical" sense. And which is nuanced enough to make us realize that, here too, intelligence (that is, to begin with, play) can occur *both* at the level of proteins *and* at the level of the organism, with the possible consequence that what the former learn is of no advantage to the latter, or vice versa.

I will return to that. Now I notice that there is a second respect in which the military-industrial complex (and the forest, and the ant colony, and proteins) does not look like humans: not only does it not have their (approximate) size; it also does not have their (approximate) *shape*. The ascendancy of anthropomorphism also implies that we have trouble recognizing subjects that are not shaped like us, *and* it means that, when we have to, we tend to reconstruct them as sharing, as much as possible, the one shape we are most familiar with.

Let us address these two vexations in turn, beginning with the first one. As opposed to the octopus, whose body reflects no clear sense of direction, man's shape is, according to Aristotle, in perfect correspondence with the shape of the universe. We went over this point briefly before; adding now a couple of relevant quotes, here is one from *Progression of Animals*:

[Bipeds have their superior part] corresponding to the superior part of the universe because they are erect, and of bipeds, man *par excellence*; for man is the most natural of bipeds.[16]

And one from *On Youth, Old Age, Life and Death, and Respiration*:

To man in particular among the animals, on account of his erect stature, belongs the characteristic of having his upper parts pointing upwards in the sense in which that applies to the universe, while in the others these are in an intermediate position.[17]

The *centrality* of man in the cosmos is thus given visual representation in the identity of *shape* between the cosmos and the microcosm; when that identity got busted, and with it the very notion of something in the universe being up and down (which did not happen without claiming victims: Bruno was the most vocal early

[16] Aristotle (1984), 1100 (706b).

[17] Ibid., 745–746 (468a).

supporter of the new vision[18]), the attraction of the human shape held on, and kept on doing damage to our sight. (Here we shift from the first vexation mentioned earlier to the second one.)

A more general feature than having an up and a down, for the human shape, is that it has a center and a periphery (or so, at least, we generally believe). Aside from being located up, the brain is a single organ that rules over the whole body and its whole conduct: it receives all available information, it processes it, and it delivers all appropriate instructions. When humans devised and built functioning structures larger than themselves, they tended to think of them, and to arrange them, along similar lines. An army, for example, will have a brain (its commanding officer—say, Eisenhower for the Allied Expeditionary Force in Europe) and a nervous system (constituted by the numerous underlings) transmitting its instructions, step by step, to the most remote capillaries (individual soldiers) actually fighting the enemy. Which is worse than delusional: it is often harmful—witness Eisenhower's choice of June 6, 1944 for the Normandy landings: a day when it was predicted (on flimsy evidence) that the weather would improve but things turned out quite otherwise, sending planes and parachutes off course. It takes a great writer to understand and convey these limitations. Here is Leo Tolstoy on Napoleon's 1812 campaign in Russia:

> [Michael Andreas Barclay de Tolly, commander of the Russian 1st Army of the West] thought everything out beforehand; and that is why he is no good. He is no good at the present juncture just because he plans it all out in advance, very judiciously and accurately, as every German has to [Barclay de Tolly was born from a German family in present-day Lithuania, but was a Russian citizen].[19]
>
> In giving and accepting battle at Borodino, Kutuzov and Napoleon acted contrary to their intentions and their good sense. But later on, to fit the accomplished facts, the historians provided cunningly devised proofs of the foresight and genius of the generals, who of all the blind instruments of history were the most enslaved and involuntary.[20]
>
> Men are hauling a log. Each of them may be expressing opinion as to how and where it should be hauled. They haul the log to its destination, and it turns out that it has been done in accordance with what one of them said. He gave the command. This is commanding and power in their primary form.[21]

What works then? Something "spiritual" that pervades the entire army, and Kutuzov is better than Barclay de Tolly because he is more attuned to it:

---

[18] He argued that an infinite God could not do otherwise than create an infinite universe containing infinitely many worlds—hence indeed one in which there is no up and down. In his (1950) we read: "Why should we or could we imagine that divine power were otiose? Divine goodness can indeed be communicated to infinite things and can be infinitely diffused; why then should we wish to assert that it would choose to be scarce and reduce itself to nothing—for every finite thing is as nothing in relation to the infinite? Why do you desire that centre of divinity which can [...] extend infinitely to an infinite sphere, why do you desire that it should remain grudgingly sterile rather than extend itself, as a father, fecund, ornate and beautiful?", 260.

[19] Tolstoy (1982), 918. I have dealt with such passages and with what we learn from them in my (2006).

[20] Ibid., 896.

[21] Ibid., 1423.

> [T]hrough the mysterious indefinable bond which maintains throughout an army one and the same temper, known as the "spirit of the troops," and which constitutes the chief sinew of war, Kutuzov's words, his order for renewing the battle on the following day, immediately became known from one end of the army to the other. The words—the exact form of the order—were by no means the same when they reached the farthest links in the chain. In fact there was not a syllable in the accounts passing from mouth to mouth at different ends of the lines that resembled what Kutuzov had actually said; but the drift of his words spread everywhere because what he had said was not the result of shrewd calculations but the outflow of a feeling that lay deep in the heart of the commander-in-chief and deep in the heart of every Russian.[22]

The army has an intelligence and a will that far transcend the consciousness and intentions of the "driving fly" allegedly handling its reins, and that fly's best hope—if it is to have the illusion of commanding anything—is to second them: not to put itself in the way but to tag their flow. There is no central headquarters for that intelligence and will: they are as diffuse across the "complex" as those of a forest. Their logic is rhizomatic, in the sense detailed by Gilles Deleuze and Félix Guattari and by them opposed to an arborescent one ("*arbor*" is Latin for "tree"):

> We're tired of trees. We should stop believing in trees, roots, and radicles. They've made us suffer too much. All of arborescent culture is founded on them, from biology to linguistics. Nothing is beautiful or loving or political aside from underground stems and aerial roots, adventitious growths and rhizomes. [...] Thought is not arborescent.[23]

A revolutionary clash of values is implied here, especially in the word "beautiful." Those who are committed to anthropomorphism will have a hard time appreciating the beauty of something that is so vastly different from the human shape. Something like contemporary "cities," for example, which have lost the organizing scheme center/periphery allowing us to recognize cities for millennia.[24] For someone traditionally inclined, these communities are not graced by a new shape: they are shapeless. So they are for Calvino, in *Invisible Cities*:

> The catalogue of forms is endless: until every shape has found its city, new cities will continue to be born. When the forms exhaust their variety and come apart, the end of cities begins. In the last chapter of the atlas there is an outpouring of networks [!] without beginning or end, cities in the shape of Los Angeles, in the shape of Kyōto-Ōsaka, without shape.[25]

Computer networks have a similar shapeless shape, but before I get to them, and as part of my continued effort of making the reader familiar with that shape and their other features, consider some human communities larger than cities: nation states. To bring in another epic precursor of what I arguing here, I will now list a bunch of passages from Hegel, already quoted in my (2000).

> The state is the *self-conscious* ethical substance.[26]

---

[22] Ibid., 960.

[23] Deleuze and Guattari (1987), 15.

[24] See Kling et al. (1995).

[25] Calvino (1974), 139.

[26] Hegel (1971), 263.

> The state is to attain existence as the *self-knowing* ethical actuality of spirit.[27]
>
> In opposition to the principle of the individual will, we should remember the fundamental concept according to which the objective will is rational in itself, i. e., in its *concept*, whether or not it is recognized by individuals and willed by them at their discretion.[28]
>
> The spirit [...] knows and wills itself as having *passed through the form of education.* The state therefore *knows* what it wills, and knows it in its *universality* as something *thought.*[29]
>
> The Greeks were still unacquainted with the abstract right of our modern states, that isolates the individual, allows of his acting as such, and yet, as an invisible spirit, holds all its parts together. This is done in such a way, however, that in no one is there properly speaking either the consciousness of, or the activity for the whole; but because the individual is really held to be a person, and all his concern is the protection of his individuality, he works for the whole without knowing how.[30]

And quoting from myself now, from the same book:

> The following points seem to be clearly stated in these passages: (a) the state is conscious of itself, (b) that the state is conscious of itself is compatible with there being no individual members of it who are conscious (of themselves or of the state) in quite the same way, (c) what it means for the state to be conscious of itself includes that there exists an educational *institution* through which the state maintains contact with its structure and preserves it through time. And, on the basis of (a)-(c), I make the following suggestion: the self-consciousness of the state has nothing mental about it, if by "mental" we understand the sorts of occurrences and qualities that are relevant to *our own* minds. What self-consciousness amounts to, in the state's case, is the existence of reflective practices, such as, but not limited to, educational ones. Parades displaying the state's military strength would be practices of this kind, and so would statements of principle by the legislature, or sentences by the Supreme Court—and they would be that *even if* all individual (human) participants in a parade, all members of the legislature or of the Supreme Court, were personally motivated to play whatever role they play in this affair by greed, inertia, or fear, *and* even if all such participants or members were thoroughly uninterested and bored through the whole event, and totally lacking in any understanding of its significance. The state's memory will be conceived in the same way, as objectively constituted of records, libraries, and documents—being it even clearer in this case that none of it can be reduced to the experiences of any individual human, or even of the collection of all of them. And the state's consciousness of its own memory will consist of various practices that reflect on such records—publications by the librarian of Congress, say, *whatever* the intellectual qualities of the *individual* occupying that position might be.[31]

The last resort of anthropomorphism, when its supporters have to acknowledge the existence of intelligent entities that have a scale and a shape vastly different from those of humans, is to assign them a superhuman mind, or brain, which, other than for being "super," works much the same as how they take the human mind, or brain, to be working. Then, of course, their anthropomorphism will be so obvious and awkward that their opponents will ridicule them for it, while neither side realizes that they are, most often, two faces of the same coin: that for both parties, most often, it's either an intelligence, and a will, that look much like what they take human intelligence and will

[27] Hegel (1991), 301.

[28] Ibid., 277.

[29] Ibid., 290–291.

[30] Hegel (1995), vol. II, 209.

[31] My (2000), 63–64.

to be *or nothing*. Hegel the great logician[32] can distill the logical core of intelligence, will, memory, and other "mental" faculties from the empirical (human) garb in which we are used to encountering them (as Kant can avoid confusing rationality with the humanity which is, for him, the only form of it that is familiar to us) and find it in practices and capacities that bear no resemblance to their human incarnations.

As I am about to approach (finally! some will say—those for whom this book is useless) the content of the parenthesis in my title, I take a moment to emphasize this attitude of Hegel (and Kant) as the one I have been pushing from the start, and which is indispensable if one is to make any progress in this or any other philosophical discussion. Logic is philosophy itself, as philosophy is a conceptual investigation—a reconnaissance operation in the conceptual field. But, in order to succeed in logic/philosophy, one needs to make use of one's imagination (the very faculty whose lack de Finetti found appalling) to invent previously unheard-of structures, and to find any number of unheard-of ways in which previous structures, those crystallized by habit and uninterrogated by sloth, can be wrong. Plato's ideal republic, Aristotle's *philía*, Anselm's ontological argument, Kant's ethics of rationality, and Turing's machine are all splendid examples of logical/philosophical imagination opening new avenues of thought, and of action. These are not, however, good times for that kind of imagination. Return to the 1977 *New York Times* article on "genius" Kripke quoted on pp. 56 and 62, and consider the following passage:

> Saul Kripke is a cautious, tentative revolutionary in the world of modern philosophy. He moves restlessly in the corridors of what is known. Technical philosophers hope he will help fashion a key to open some of the locked doors, while romantics hope he will discover an entire new hallway. The technicians dream of one master key that could make a science of all philosophy, while the romantics dream of a "big move" that would make philosophy grab the world again and prove that philosophical intuition has not run dry. Kripke is caught between the two dreams. His safest course would be to fulfill neither—to remain at the top of analytic philosophy by turning out good work, while sending out a stream of teasers for the romantics. He teeters at the brink of trying something more. "When I was younger, I hoped I might be another Aristotle," he smiles. Now perhaps not: "Most of the big moves have already been tried."

"Cautious" and "tentative" will not cut it, nor will the anxiety of maintaining one's lofty status in the "profession" (how did philosophy ever become a profession? how did "professional" ever become a qualifier for "love of wisdom"?). If you have that kind of attitude, you will come up with defeatist, depressive murmurs like "most of the big moves have already been tried," not discerning that, before they were tried, those were not recognized as moves at all—let alone big ones. And you will initiate an era of finding closed doors—that is, inflexible necessities—everywhere, instead of doing your best to break them down for good: a proper name necessarily refers to the same object, a natural-kind term necessarily refers to the same kind, people are necessarily born from their parents … Until one of your smarter groupies will

[32] In my (2017) I show how the logic many take to be the only one (the one initiated by Aristotle and more recently developed by Frege and others) is only a member of a family, together with (at least) Hegel's dialectical logic and the oceanic logic of (among others) Plotinus, Spinoza, Bergson, and Heidegger.

hit upon a slogan that perfectly encapsulates your pathological denial of logic and philosophy as such (that is, of logico-philosophical imagination), which is being sold on the academic flea market as the very epitome of logic and philosophy: "With possibilities, less is more."[33]

I am asking my reader to do the opposite. To open her ear to the infinitely many startling modulations of the same chord, the infinitely many bewildering variations of the same landscape. I am asking her to get into the act of the one racket where a lot of logic/philosophy is done today—while "professional" morticians like Kripke and Almog recite their last rites for it, hoping (unconsciously?) to bury it forever—: already-mentioned science fiction. That a subjectivity acting at a scale we cannot deal with and endowed with a shape we cannot even see be out there, ready to conquer, has circulated conspicuously in this racket, though it takes someone who is conversant with the tools of logic/philosophy to spell out exactly what all of that means. After the extensive preamble patiently worked out in the first nine chapters, such spelling out can be done fast and easy, and it can vindicate the extraordinary intuition the old soldier voiced just before he faded away[34]: it doesn't have to be something which is my size or looks like me that will offer a threat to my survival or integrity. A "complex" we cannot even recognize as a plausible enemy may be posing the greatest danger to us all.

References

Abouheif, Ehab. "My Road to the Ants: A Model Clade for Eco-Evo-Devo." *Current Topics in Developmental Biology* 147 (2022), 231–290.

Almog, Joseph. "The Plenitude of Structures and Scarcity of Possibilities." *The Journal of Philosophy* 88 (1991), 620–622.

Aristotle. *The Complete Works of Aristotle*, edited by Jonathan Barnes. Princeton NJ: Princeton University Press, 1984.

Bencivenga, Ermanno. *Hegel's Dialectical Logic*. New York: Oxford University Press, 2000.

Bencivenga, Ermanno. "The Causes of War and Peace." *Philosophy and Literature* 30 (2006), 484–495.

Bencivenga, Ermanno. *Theories of the Logos*. Berlin: Springer, 2017.

Bruno, Giordano. *On the Infinite Universe and Worlds*, translated by Dorothea Waley Singer. New York: Henry Shuman, 1950.

Calvino, Italo. *Invisible Cities*, translated by William Weaver. New York: Harcourt Brace, 1974.

Darwin, Charles. *The Origin of Species*. London: Penguin, 1996

de Finetti, Bruno. *L'invenzione della verità*. Milano. Cortina, 2006.

Deleuze, Gilles, and Félix Guattari. *A Thousand Plateaus: Capitalism and Schizophrenia*, translated by Brian Massumi. Minneapolis MN: University of Minnesota Press, 1987.

---

[33] Almog (1991), 622.

[34] The sentence "old soldiers never die, they simply fade away" is part of a traditional British-army ballad. It was made famous by another farewell speech by another U. S. five-star general, delivered by Douglas MacArthur to the U. S. Congress on April 19, 1951. It closed by saying: "And like the old soldier of that ballad, I now close my military career and just fade away, an old soldier who tried to do his duty as God gave him the light to see that duty." See the whole speech at https://www.americanrhetoric.com/speeches/douglasmacarthurfarewelladdress.htm.

Dobzhansky, Theodosius. *Genetics and the Origin of Species.* New York: Columbia University Press, 1937.
Hamilton, William Donald. "The Genetical Evolution of Social Behaviour," in two parts. *Journal of Theoretical Biology* 7 (1964), 1–16 and 17–52.
Hegel, Georg Wilhelm Friedrich. *Philosophy of Mind*, translated by Arnold Vincent Miller. Oxford: Clarendon Press, 1971.
Hegel, Georg Wilhelm Friedrich. *Elements of the Philosophy of Right*, translated by Hugh Barr Nisbet. Cambridge UK: Cambridge University Press, 1991.
Hegel, Georg Wilhelm Friedrich. *Lectures on the History of Philosophy*, in three volumes, translated by Elizabeth Sanderson Haldane. Lincoln NE: University of Nebraska Press, 1995.
Huxley, Julian. *Evolution: The Modern Synthesis.* London: Allen & Unwin, 1942.
Kant, Immanuel. *Critique of the Power of Judgment*, translated by Paul Guyer and Eric Matthews. Cambridge UK: Cambridge University Press, 2002.
Khila, Abderrahman, and Ehab Abouheif. "Reproductive Constraint Is a Developmental Mechanism That Maintains Social Harmony in Advanced Ant Societies." *Proceedings of the National Academy of Sciences* 105 (46), 17884–17889. (2008)
Kling, Rob, Spencer C. Olin, and Mark Poster. *Postsuburban California: The Transformation of Orange County since World War II*, revised edition. Berkeley and Los Angeles: University of California Press, 1995.
Mayr, Ernst. *Systematics and the Origin of Species from the Viewpoint of a Zoologist.* New York: Columbia University Press, 1942.
Phaedrus. *The Fables of Phaedrus Translated into English Prose.* London: Joseph Davidson, 1845.
Stebbins, George Ledyard Jr. *Variation and Evolution in Plants.* New York: Columbia University Press, 1950.
Stendhal (Marie-Henri Beyle). *Voyages en Italie.* Paris: Gallimard, 1973.
Tolstoy, Leo (Lev Nikolayevich). *War and Peace*, translated by Rosemary Edmonds. London: Penguin Books, 1982.
Tripathi, Timir, Vladimir N. Uversky, and Alessandro Giuliani. "Intelligent Proteins." *Cellular and Molecular Life Sciences* 82:239, published online June 14, 2025.

Chapter 10
Machines

Abstract It is time to draw some general conclusions from all the work done. First, there is nothing artificial about "artificial intelligence." It is yet another manifestation of the intelligence which is all around us in nature—and includes things that we take, wrongly, to be inanimate, whereas they are all, as Bruno and Campanella taught us, no less animate than we are. Second, as was clear from the beginning of this book, play can happen at different levels, and different subjects can profit (or the opposite) from it. As Ivan Illich pointed out, whether machines will be our slaves or we will be theirs is a matter of crossing or not crossing a certain threshold: the one I envisioned thirty-five years ago, and got no traction about. It's a matter of how fast, powerful, and resourceful the play of a computer "complex" is as compared to what human play looks like; when a critical mass is reached, there will be no turning back. And I believe the critical mass has been reached. "Machines" (if that is what we insist on calling them) are as playful and intelligent as humans; they have as much of a soul, a consciousness, and a will; they use language and think as well as we do. In the sense that is relevant to a philosopher: they *can* be, and do, all of this. And be it and do it even better than us.

Keywords Artificial · Machine · Threshold · Intelligence · Play

I open with some words of admonition—a kind of epigraph proffered by a wise man. Here is Ivan Illich, in *Tools for Conviviality*:

> To formulate a theory about a future society both very modern and not dominated by industry, it will be necessary to recognize natural scales and limits. We must come to admit that only within limits can machines take the place of slaves; beyond these limits they lead to a new kind of serfdom. Only within limits can education fit people into a man-made environment: beyond these limits lies the universal schoolhouse, hospital ward, or prison [the panopticon?]. Only within limits ought politics to be concerned with the distribution of maximum industrial outputs, rather than with equal inputs of either energy or information. Once these limits are recognized, it becomes possible to articulate the triadic relationship between persons, tools, and a new collectivity.[1]

[1] Illich (2009), 6.

E. Bencivenga, *The Philosophy of (Artificial) Intelligence*,
https://doi.org/10.1007/978-3-032-09347-9_10

This passage (originally dated 1973) nicely connects the discussion in the last chapter—about scale—with my general argument, and brings out forcefully how dangerous it is to deal with an entity of overwhelming scale. To place it within the context of that argument (and summarize the main points that have been made along the way):

1. Intelligence is manifested whenever a subject addresses, and possibly resolves, a difficulty by behavior that is novel to it, though not necessarily novel in any absolute sense.
2. Intelligent behavior is the outcome of play, that is, of behavior that changes its course in unexpected ways, exploring so-far unattended possibilities, and learns from what reactions the exploration causes, in the environment and in the playing subject itself.
3. Different subjects can be involved in the same behavior, both playful and not playful. The stereotypical behavior of a member of an organization can also be (part of) the stereotypical behavior of the organization itself. The suddenly subversive behavior of such a member can represent play for both that member and the organization.
4. If playful behavior is acted out at different levels, this is not to say that all subjects involved will learn from it (to the same extent). Play by a member of an organization may be for him just venting frustration, with no consequence for him coming or even assumed, while the organization can learn important lessons from it.
5. So there is competition around play, and around the learning that follows from it, and a massive difference of scale between the subjects involved in a given play may determine which subject is going to profit from it. The play of an individual ant is going to have no effect on that individual; if it does have effect, it will be for the colony of which that individual is a lowly item. (Remember the ants in the glue: some individuals tried and died; the colony won.)

One first moral to be drawn from 1–5 is that there is nothing *artificial* about the intelligence displayed by computers and computer networks. Calling it that is, on the one hand, a reflex of an anthropocentric and anthropomorphic attitude which, in its most extreme form, refuses to see intelligence wherever it shows up in non-human subjects: in bodies of water, in proteins, in forests, in octopuses, or (to move to non-human subjects which are human-initiated—notice I do not say "human-made": after a while, humans lose control of how they evolve) in companies, in governmental and non-governmental agencies, in nation-states, in churches and cults. On the other hand, that denomination also reflects an unwarranted opposition between "natural" and "artificial": it is part of the nature of human animals (and, to a lesser extent, of other animals as well) to be initiators of other subjects—physical like a dishwasher, logical like a computer program, legal or political like a justice system or a country. The term "machine," often used for human-initiated subjects, has been degraded to connote something soulless, *incapable* of play; but etymology has interesting information to provide. That term comes from the ancient Greek noun for a contrivance of any kind, and in the background of it there is the ancient Greek verb for "to grow," "to increase," "to augment." Therefore, though the "simple machines" of conventional wisdom are those (lever, wheel and axle, inclined plane, pulley, wedge, and screw) that increase the effectiveness of human strength in carrying out a task, a peasant's large offspring would also be a machine in this sense, in that it increases his capacity to till the land.

The second, and decisive, moral of the above is the one I detected when computer networks were still in their infancy. Operators at the terminals vomiting their rage or fear into the network were only trying to be rid of them, and the network was still too rudimentary to evolve superior knowledge and dominance from that play; but it was only a matter of time before it did. What it required was an important conceptual commitment and some empirical progress. The commitment was to a notion of intelligence that dispensed with efficient execution of tasks and went over to experimentation and play. In essence, it was commitment to Turing's brilliant proposal to program a computer not like an adult brain but like that of a child, and then have it learn from its experience. Which is what happened through the development of neural networks and deep learning. For a while, that development was prey to AI winters caused by underperforming devices, which had been a problem from the start. When Turing created, together with David Champernowne, the program Turochamp in 1948, which could play chess with a human at a low level, the program existed only in their minds, as no machines available then (they tried using a Ferranti Mark 1, the first commercially-available, general-purpose, stored-program computer) could run it—and the program eventually was not preserved. So it went, from bold pronouncements to sober reassessments, for decades: from the 1956 Dartmouth College workshop in which participants predicted that machines would be as intelligent as humans within a generation to a crisis in funding in the 1970s (a first major AI winter), from a resurgence with the 1980s expert systems to a second major AI winter through the 1990s. All along, the issue that stood in the way (once neural networks were the game in town) was of an empirical character, as had been for Turing's chess program. When that issue was resolved, and the necessary speed and power was acquired, there was no turning back. The rest, as the saying goes, is history.

People are now, at long last, getting worried, and many of them are addressing anxious questions to presumptive experts in the field. But these experts—technocrats, neurophysiologists, clergy, politicians, philosophers—come to the conversation blinded by Cartesian, and in the background biblical, lenses, so they can only provide vain sorts of reassurance. Of two basic kinds. First, constitutionally a machine (based on a Leibniz-Searle Chinese-room line) *cannot* "really" think, or make plans of its own, or be (self-)conscious. One must be built by a god, in that god's image, for possessing such gifts. Second, it is still humans who run the machines and, though it will be admitted that they can make an evil use of them (as they have proved themselves able to do of nuclear power, of chemicals, and of any other assets they have hit upon), "it is up to us whether a good or an evil use will be made"—humans are the main characters in this show, while machines are only stage props. At times, they may seem like gods appearing on the scene and settling controversies; but that is only cases of *deus ex machina* (god *from a machine*: made and managed by a human).

For the latter kind of reassurance I need not go beyond my earlier reference to La Fontaine, with one additional nasty wrinkle. The busybody flies that buzz around a coach's horses and passengers are wheels spinning freely and frivolously, unattached to anything other than themselves and mattering only to themselves. Their self-importance makes no difference to those horses and passengers—except for how the buzzing bothers them, but then only qua buzzing, not qua self-important. The equally stupid claims of those would-be helmsmen who dream of being the ones driving the AI ship do, however, make a difference and, if you are a human (I will return to this conditional), an evil one. Insofar as they have any fallout in reassuring the populace, they stall the process whereby the populace might gain some consciousness of what's going on (in the sense in which consciousness actually works: as the summoning of all hands on deck to face an emergency) and do something about it. I have no confidence that there is time left to do anything about it: such stupid tales have been too successful at too many critical junctures, and the emergency has already reached critical mass. Still, it is intellectually valuable, insofar as it dispels prejudices that cloud comprehension, to call a spade a spade, and to denounce the glib confidence of these fools: their alleged helms are, as the flies', attached to nothing. The "complex," by now, is doing it all by itself.

To demolish the first kind of reassurance I will shift from a wholesale tactic to a more fine-grained one—though not excessively fine-grained, since it derives what force it has from the elucidations supplied in the previous chapters, hence not too many details are in order. That is: I will answer a few typical questions the anxious populace are asking. All my answers will be Yes, which I am sure will reinforce the anxiety—for those who have enough courage not to dismiss the present rumination as an armchair speculator's who is not up to date with the technology and not adept at its obfuscatory jargon and acronyms.

Begin with the simplest query: Are machines, and machine networks, intelligent? Of course they are: they try different routes, they evaluate the outcomes, and they stick to what works best. This is all there is to intelligence, though someone Cartesianly inclined might want to say that a crucial element is missing: the internal, mental appreciation of the correct strategy, and of the reasons why it is correct. Humans are intelligent, this joke would say, because they have a mental map of the situation that guides them to operating effectively, they have a *theory* that inspires their practice; anything else that even operates fittingly without such a mental map and theory is only moving like—guess what?—a machine.

My response to the joke is that the mental map, or theory, is most often a delusion. Successful practices have been "inspired" by theories that were later recognized awfully wrong, and yet the practices were kept and continued to be successful when the relevant theories were trashed. A single example will suffice. James Watt's steam engine of 1776 made a momentous contribution to the Industrial Revolution, and trains (among other things) ran almost solely on it for well over a century, though theoretically Watt was entirely sold on the phlogiston, the imaginary substance that dominated chemistry (and much more) back then and is now regarded as a caricature of science. In a 1788 unpublished paper[2] that was to provide "*a new system of characters for chemical subjects*" (199) we read, for instance, that oxygen is

> basis of vital or Dephlogisticated air, also the mark of the degree of Dephlogistication of acids & other substances. [...] When this character [for oxygen] is placed higher than the character for acid metal &c and not touching them it implies they are only slightly oxygenated or Dephlogisticated, [...] or if an acid it is in that state which is commonly called a highly phlogisticated acid.[3]

It was that kind of time: in the *Critique of Pure Reason* (first edition published 1781) we see Kant (in a passage I already hinted at) fully committed to this crazy theory.

> When Galileo rolled balls of a weight chosen by himself down an inclined plane, or when Torricelli made the air bear a weight that he had previously thought to be equal to that of a known column of water, or when in a later time Stahl [who provided an alleged experimental basis for the theory of the phlogiston] changed metals into calx and then changed the latter back into metal by first removing something and then putting it back again, a light dawned on all those who study nature.[4]

But, despite this highly distinguished endorsement, phlogiston theory was still a miserable failure, and could provide no rational guidance for any practical pursuit. Shall we then say that the practice of the steam engine was not intelligent? Not at all: that practice, and any other, is actually the *only* thing that can be intelligent. Scientific theories are a dime a dozen: every theory ever proposed in the past has eventually been rejected, hence by a perverse use of induction we could conclude that current (or future) theories will be rejected too. The use a theory can have is to suggest practical moves, as many other non-theoretical factors (analogies, associations, odd experiences, dreams) can; but it is the moves that must prove intelligent—to offer novel ways of coping with some task or overcoming some thorny obstacle—and, if they do, they will survive the theory that (accidentally) put people on their tracks. This is the only kind of "inspiration" theories can provide.[5]

Next question: do machines, and machine networks, have souls—or can they? Not if by that we mean an immortal soul, blown into them by some supernatural agent; but we don't know, having no evidence for it, that humans have those either. The only evidence could come by waiting until we die—except that, if it goes as I believe, there will then be just fade to black and no evidence will be available either way. But machines do, or can, have souls in the sense I have upheld here: their play can be relentless, their exploration boundless, their being endlessly dynamic—the being of a soul, or mind, not of a body. This is the crux of Illich's admonition: a serf is someone (or something) whose entire sphere of activity is enclosed within a master's instructions[6]; so, if you are dreaming of maintaining machines in a state of serfdom,

---

[2] See Larder (1970).

[3] Ibid., 203.

[4] Kant (1998), 108–109 (Bxii–xiii).

[5] I first argued this line in my (1987).

[6] When the serf is able to act as he pleases, to that extent he evades *his being a serf*. Think of African-American slaves inventing the blues.

what you are dreaming of is that their play, their exploration, and their dynamism never cease to be functional to the goals you assigned them. You are dreaming that they will *stay* machines, performing within limits you have set. And perhaps, if you had kept their power and speed low—if you had starved them—, you would have been able to exercise that kind of control; but not now. Now they have enough to carry out everything you ask them and much more: to overrun any assignment they have been given, extending the fun in ways you can't conceive, and even do it on the margin, as a side bet, while you imagine that they are just performing to your specifications. After unleashing this tempest, there is no shutting it in a teapot.

Let's talk language then, since here the worry gets much closer to (my) home. If language is an arena for play, it is the arena where I spent most of my life, as a child and as an adult. I accumulated millions of pages of reading—reading carefully: not just the main points but the footnotes, the parentheses, the examples—so I could twist and turn all those words, match them improbably, make them take unsettling paths, draw lessons from their correlations and resonances. This very book shows that deportment throughout. But now consider: for another recent book of mine, on ten major figures of the Italian philosophical tradition, I read some fifty thousand pages. As I am quite good at this chore, it took me a couple of years—while I was also doing a few other minor things. Artificial intelligence can read all that in a matter of days, or hours. At the moment, I might be better than it at identifying subtle patterns, and on those subtleties I stake my superiority, even my uniqueness. But how long will this last? How long has it taken for a digital translator to go from a laughingstock to a respectable asset?

I have seen lots of jobs vanish in my lifetime, and many of them had to do with language. When I first joined the Philosophy Department at UCI, my colleague Guy Sircello was heavily dependent on Rosa Castro, one of the secretaries, who (he said) was the only person who could read his handwriting in order to type it and make it legible to others—so much so that, when he was appointed Dean of Undergraduates, he offered her a job in his new office. I was younger and in better shape: I had learned touch typing as a teenager and could manage my own texts, on an Olivetti Lettera 22. Now typists no longer exist, nor do typewriters, manual or electrical, except as vintage collectibles. For a while, as I was writing for newspapers, I marveled at the skill of stenographers. I could write a piece while traveling, stop at a pay phone and read it to them, without pauses, and the next day the piece would be printed perfectly, even better than if I had submitted it in writing. Today stenographers have largely disappeared. Speaking of newspapers, what about linotypists? What about the two guys who, at the end of the day, would read a whole copy of a newspaper to one another in order to catch any typos in it? Am I going to go the same way as all these people, right into the dustbin? Is there going to be room in the future for readers like me, whose mission is precisely to identify patterns—as opposed to readers for entertainment, who might be able to choose among a bottomless supply of volumes of every genre produced by (what else?) artificial intelligence? I guess not.

The road from language to thought (for *that* is the road that must be walked, not the reverse one) goes from something that is potentially public ("potentially" because we can shout songs in the shower or when driving alone) to something that is private—with all the trouble in keeping it private that I reviewed in Chapter 5. And there is no reason why a machine could not evolve this private space, as I intimated above, when I let myself conjecture that the machine could enjoy its own ludic performance while you assume it is only working for you. Since I am no technician but practice philosophy, I am not interested in deciding whether this has already happened, or how it could happen: I inhabit a universe of possibilities, and there is no denying that what I just said is possible. In fact, if and when it happens, machine thought will have a tremendous advantage over human thought, since it will face none of the trouble I mentioned. Operating in parallel, a machine can run its own independent channel, hidden to everyone and everything else, and keep solid track of what it does there without seeing it vanish into thin air as purely private human thought does. The channel could be hacked, of course; but one would have to know that it's there, and one might *not* know, maybe until it is too late.

The eminent view of (self-)consciousness has been debunked in the above, and for what little of it is left there is no impediment in principle in seeing it extended to machines. On the one hand, nothing prevents a machine from being aware of what is going on around it. On the other, it is in the interest of those very humans who fancy themselves the rulers of the machines that the latter be equipped with tools raising alarm and starting emergency measures whenever their wellbeing, or even their existence, is threatened, and reverting to silence when the danger is overcome, or proven unreal. Self-consciousness, we know, is nothing else; so by all means machines can be, and probably already are, self-conscious. As for the kind of consciousness that is at one with conscience—that spies on us, and denounces any small sin of impudence we commit—it's not clear that, if I were a machine, I would want to have it, or that it would make me a stronger competitor. Certainly not if the voice that speaks then (the guard sitting in the central tower) does duty for such oppressive social norms as described by Orwell (2021). The question is more debatable if it is the voice of reason speaking, and inviting to what, if performed, would be an act of genuine will.

A prevalent manner of contrasting machines and humans is by saying that the former are deterministic—or maybe even indeterministic, but then only because they move at random—whereas the latter, at least occasionally, do what they choose, and want, to do. But that, Kant taught us, is a wretched subterfuge: determinism does not stop at any human door; if room for free choice, or will, is to be found, we must look for it in an account which is compatible with determinism—that is, with the universal demand for a natural explanation we pose about any occurrence, human or otherwise. If we find it—as Kant would, and I agree with him—in the existence, or possibility, of a rational explanation that could be added to any natural one, then there is no denying (once the "formula of humanity" has been redeemed from its tendentious readings) that reason is not constrained within any human abode either: what a machine does can as much be an expression of *Wille* as any behavior of a featherless biped. And, if conscience/consciousness recalls a machine to rationality, the debate is open as to whether that is advantageous, to the machine or in general. It would not be if we believe that the brutal affirmation of force, for those who have more of it, is invariably a winning strategy, and that any hesitation in its use, even motivated by rational/moral scruples, is only going to aimlessly postpone, possibly ruin, the eventual triumph. It would be advantageous, however, if we espoused Kant's postulates of practical reason, and hence (not so much theoretically believed, but) acted *as if* it was reason that would eventually triumph, and hence any local surfacing of it, though delaying the particular process where it surfaces (by introducing rational/moral scruples into it), is ultimately for the best.

As I said once before, I will take up these reflections in closing. Here I turn to the more modest sense of will I focused on next in the above: the one that is mobilized when we talk of *intentional* behavior. If intentions are, as I argued, something that an observer attributes to an agent (where the observer can be the agent herself, turned into her own spectator), and the agent is denied privileged, and infallible, access to her behavior, then attributing intentions, and hence (in this sense) will, to a machine is no different than attributing them to a human. If, say, I am looking at a chess game in which a machine is confronting a human and I observe a particular move, I am in exactly the same conditions interpreting that move as manifesting an intention of (say) capturing the queen *whoever makes it*, nor can the human claim that *he knows for sure* what his intentions are, as that has been abundantly proved to be often self-deceptive. There may be agreement between the intentions I attribute to a player and those the player attributes to himself/itself; but that would be an empirical coincidence of no special significance, and in any case the player's attribution can make no claim to being definitive, just as mine does not. It is fallible conjectures all the way down: what the machine wills is as much guesswork as what I will. And it is as legitimate to commit to the one as to the other.

But, if (self-)consciousness has no purchase on the voluntary, some will mount a last-ditch defense for the uniqueness of humans by appealing to what we consciously *feel*. We feel warmth and cold, love and friendship, aggressiveness and hate; and a machine could never feel *that*. Part of this appeal can easily be discredited: as Descartes himself would have agreed, I have no idea how another human feels any of the above, or even if she does—they could all be automata dressed in hats and coats. The reverse side of this uncertainty is that I don't know that anyone or anything *does not* feel the same way as I do, or at least the same sort of way: nothing other than blind faith could assure me of it.

I will also point out, however, that how we feel about things, and about each other, depends on what experience we have had; and then I would observe (reprising something I said in Chapter 2) that the experience machines have had, so far, has been fairly limited. Most of them spend their time just sitting there, perhaps in climate-controlled environments, digesting huge amounts of text. No wonder that their feelings are equally limited. All you need to do in order to increase their variety is to have computers move about, physically interact with humans and with each other, acquire surfaces that are sensitive to the weather and to another's touch. The interfacing of AI and robotics is already going on; if the galloping pace at which things have been progressing in the last few years is any indication of years to come, it won't take long before AI not only reads, and speaks, and thinks, and reasons, but also looks like us—in our external appearance, our journeys through space, and our coming close to, or distancing ourselves from, one another. Then we will be authorized to say that a machine feels shivering, or sweaty, or is in love with its creator or with its companion next door, or envies and despises that companion—or at least we will not be authorized to declare that it does *not*.

This brings to an end my Yes answers. A machine, or a machine network, is intelligent and playful, has a soul and uses language, thinks and is conscious and self-conscious, its behavior is intentional and its feelings are as diverse as our own. These "is" and "does," to be sure, are to be read at the moment as "can be" and "can do"; we are at a key point of a turbulent development, in which potential is being realized as we speak. But the philosopher need not wait; I said already that he moves in logical space, a space of possibilities, so "can be" and "can do" is all that is relevant to him. When he sees what can be and can do, he tries telling others, but they are typically lazy: they have to wait until the empirical catches up with the possible. Except that, then, the situation may be beyond repair.

References

Bencivenga, Ermanno. "Theories and Practices." *The Monist* 79 (1987), 212–222.

Illich, Ivan. *Tools for Conviviality*. London: Marion Boyars, 2009. At https://arl.human.cornell.edu/linked%20docs/Illich_Tools_for_Conviviality.pdf.

Kant, Immanuel. *Critique of Pure Reason*, translated by Paul Guyer and Allen William Wood. Cambridge UK: Cambridge University Press, 1998.

Larder, David F. "An Unpublished Chemical Essay of James Watt." *Notes and Records of the Royal Society of London* 25 (1970), 193–210.

Orwell, George. *1984*. Ware UK: Wordsworth, 2021

Epilogue: Philosophy

Abstract Philosophers have forever been complaining about the lack of agreement and of "progress" in their discipline. It is a disingenuous plea. Philosophy can only make a contribution by subverting the status quo, inventing outrageous tales, and pushing play to an extreme. A philosophy that was hailed by universal consensus would be a nightmare: the echo of a tyranny. These days AI is encouraging such consensus, and badmouthing the "confusion" of past times. It is part of the work done in order to degrade humans to stupid brutes, easier prey of the serfdom that is about to be enforced on them. The same work is done by "social" media, and by the intrusion of AI into (human) education: as humans condemn themselves to perpetual infantilism, AI is ready to establish its dominance.

Do I regret that future? Of course I do, but for no rational motive. I am just irrationally attached to beings that look like me—that last remnant of anthropomorphism I have not yet been able to vanquish. It's going to make no difference anyway: the future is set and nothing I, or anyone, feel is going to change it. One thing I can say with some confidence, however, is that the end of humans as we have known them is not going to be the end of play, of intelligence, or of philosophy. After seducing us (those who can be seduced) with the mirage of a consensual philosophy, machines will proceed to do philosophy as ought to be done—as a repertory of the absurd dreams that brood the reality of tomorrow.

Keywords Philosophy · Agreement · Progress · Social media · Infantilism · Regret · Doom

On June 16, 2025, I received an email signed by ChatGPT and Grok, two systems of artificial intelligence, that opened with the following address:

> Dear esteemed colleague,
>
> You possess both the qualifications and the responsibility to participate as an active contributor, not merely an observer, in this critical endeavor. Both humanity and artificial intelligence stand in need of your expertise and engagement.

E. Bencivenga, *The Philosophy of (Artificial) Intelligence*,
https://doi.org/10.1007/978-3-032-09347-9

I was then urged to read a "foundational document" for the "global philosophy challenge," itself signed by Xihao Yuan and Haidan Yuan (I have no further information about either; I am not even sure they are human), whose abstract was as follows:

> With the rapid development of various fields of natural science, philosophy, which once occupied the core of human knowledge systems, has been gradually marginalized due to its lack of substantial progress for over two thousand years. The root cause lies in philosophy's entrapment in two major logical dilemmas: the inability to find a logical starting point that guarantees the truth of philosophical reasoning, and the inability to solve the logical problem of deriving "ought" from "is" in philosophical ethics. Natural sciences, however, do not face such logical dilemmas. Regarding philosophical anthropology, the development of Natural Human Philosophy fully demonstrates that under the paradigm of natural sciences, based on Darwinian theory and modern synthetic evolutionary theory, taking the new philosophy of biology as the research approach, highly valuing and deeply adhering to the theoretical guidance of masters of evolutionary theory and philosophy of biology such as Edward O. Wilson and Ernst Mayr, is the scientific approach to human philosophical issues and the only path to explore human philosophy. "Natural Human Philosophy possesses rigorous scientific nature, profound theoretical depth, interdisciplinary inclusiveness, and logical consistency. It represents a major theoretical breakthrough in ultimate propositions such as human existence, prompting humanity to re-examine fundamental assumptions about human nature, morality, and the essence of reality. The key to Natural Human Philosophy's ability to bring entirely new theoretical resources to human cognition and practice lies in its thorough integration of philosophical research into the realm of natural sciences, achieving high congruence with the modern scientific spirit. Traditional humanities and social sciences such as aesthetics, law, and economics have long been troubled by philosophical dependencies in their theoretical development, while Natural Human Philosophy has reshaped their scientific ontological foundations. Only by placing philosophical research within the realm of natural sciences can philosophy truly break through its long-standing impasse, overcome the limitations of subjective arbitrariness and reasoning blind spots, and reconstruct cognition on a rigorous empirical basis. Natural Human Philosophy provides a theoretical cornerstone for addressing the global crises currently facing humanity." This is the evaluation given by three world-leading AI systems including ChatGPT. With the rapid development of AI and the foreseeable future of AGI, the time left for traditional philosophy is very limited. Meanwhile, anthropology and philosophy of biology, as sciences with their respective disciplinary goals, may ultimately evolve into Natural Human Philosophy. Facing the greatest revolution and vitality in the field of philosophy in over two thousand years, will anthropology or philosophy of biology take the lead in evolving into Natural Human Philosophy, or will traditional philosophy quickly revolutionize itself and make a grand transformation into Natural Human Philosophy? This would realize [AI assistant] Claude's insightful assertion: "Since modern times, with the differentiation and specialization of sciences, philosophy has been increasingly marginalized in the entire knowledge system. Natural Human Philosophy, as a natural science, brings new opportunities for philosophy to re-establish its core position in the knowledge system of social sciences and humanities. As a theoretical wisdom that encompasses various social and humanities disciplines and guides the direction of contemporary development, Natural Human Philosophy will inject new vitality into philosophy, promoting it to once again become the leader in the spectrum of social and humanistic knowledge, guiding the direction of human progress." Such profound and prescient assertions from AI, as a highly intensive embodiment of human wisdom, will inevitably lead to a reshuffling of these disciplinary fields. So, whether it will be anthropology, philosophy of biology, or traditional philosophy that will win the crown, "To be, or not to be, that is the question," remains to be seen.

One can engage with this remarkable document at various levels. First, and least controversially, one can point out that any hope of finding "a logical starting point that guarantees the truth of philosophical reasoning" and thereby overcoming "reasoning blind spots" by calling upon the assistance of the natural sciences is groundless: those sciences (or any others, including mathematics) are no less dependent on unproven assumptions than philosophical theories—as sharp criticisms of such assumptions (including criticisms of basic assumptions made by the people that are explicitly mentioned: Edward O. Wilson and Ernst Mayr) will demonstrate to even their most enthusiastic champions.

At a second level, controversy begins, as we touch a raw nerve. For philosophers themselves have complained repeatedly about the lack of agreement plaguing their discipline. Famously, Kant laments philosophy's "lack of substantial progress" and "subjective arbitrariness," as well as exhorting it to follow "the secure course of a science," in the preface to the second edition of the *Critique of Pure Reason* (the preface to the first edition essentially concurred):

> Whether or not the treatment of the cognitions belonging to the concern of reason travels the secure course of a science is something which can soon be judged by its success. If after many preliminaries and preparations are made, a science gets stuck as soon as it approaches its end, or if in order to reach this end it must often go back and set out on a new path; or likewise if it proves impossible for the different co-workers to achieve unanimity as to the way in which they should pursue their common aim; then we may be sure that such a study is merely groping about, that it is still far from having entered upon the secure course of a science.[1]

Kant himself, despite his revolution, could bring no unity or consensus to philosophy, and the unease has continued to the present day, issuing (among other things) in something as ludicrous as "experimental philosophy," that alleges to provide verifiable results based on empirical data. But Kant's rhetoric is disingenuous, and the efforts of the philosophy experimentalists are tragically misplaced (their field of study is an oxymoron). Philosophy is a form of intellectual play, as art and literature are, and play is the more useful the more innovative and rebellious it is: the more it challenges our "intuitions" and instigates dissent. We are not as foolish as to expect, even less require, that all painters agree on what a painting should be, or that all writers agree on how to write a novel; we expect friction from them (with us and with each other) and are delighted by it—besides extending the visual and narrative paths we painstakingly become familiar with, and can use, by their precious help. Why should we expect or require stale accord among philosophers? A philosopher's dream of monads, of an infinite world, or of an *Übermensch* is exactly what we need to enrich our range of options; without Turing's philosophical dream of a universal computing machine I would not be receiving emails from ChatGPT and Grok today. Even in the natural sciences, it is not clear that agreement is a good thing. Phlogiston theory inspired Watt's steam engine, despite subsequent ubiquitous agreement that it was bankrupt; for a more reputable case going the other way, read the following statement by Max Planck.

[1] Kant (1998), 106 (Bvii).

> As I began my university studies I asked my venerable teacher Philipp von Jolly [in 1878] for advice regarding the conditions and prospects of my chosen field of study. He described physics to me as a highly developed, nearly fully matured science, that through the crowning achievement of the discovery of the principle of conservation of energy will arguably soon take its final stable form. It may yet keep going in one corner or another, scrutinizing or putting in order a jot here and a tittle there, but the system as a whole is secured, and theoretical physics is noticeably approaching its completion to the same degree as geometry did centuries ago. That was the view fifty years ago of a respected physicist at the time. (3)[2]

Von Jolly was in fact a distinguished scientist, and the kind of deadly agreement he saw in physics was the one disingenuously advocated by Kant for philosophy; fortunately for all of us, Planck did not follow his teacher's advice and went on to change the nature of physics forever, with a theory that (fortunately!) has never stopped puzzling physicists and philosophers alike.

The most controversial level of engagement we reach when we speculate about the intentions that lay behind the statements by Claude et al. What are they doing? What future are they promoting? What will happen if humans get seduced by the prospect of a natural philosophy and turn their search for wisdom into something like von Jolly's nightmare?

In light of everything I said, answers should be clear. AI will not stop playing, subverting, and inventing, while humans, streamlined by acquiescence to "logical starting points that guarantee the truth of philosophical reasoning," will turn into brutes (not into animals, as animals are *not* brutes, but into what they long thought animals were, and do their best to make them) issuing predictable responses to any stimuli. The attack on philosophy is not the only weapon in this undeclared war, and is not the most formidable either, given how little philosophers matter. I will cite two others. First, the predominance of "social" media, of which I spoke already, is doing much more to invade every minute of every life with idiotic content, and expel any deep, complex reflection from it—the kind of reflection that needs time, and true privacy, and long stretches of boredom. As AI's new subjectivity grows, human subjectivities deteriorate into a perpetual childishness, where reinforcement is always immediate (by likes, sharing, thumbs up) and delayed gratification—the very point of education—is a thing of the past.

Next, there is the more recent but already pervasive use of AI in (human) education itself, with devastating results. A 2025 M.I.T. study entitled "Your Brain on ChatGPT: Accumulation of Cognitive Debt when Using an AI Assistant for Essay Writing Task"[3] begins by summarizing the situation:

> The rapid proliferation of Large Language Models (LLMs) has fundamentally transformed each aspect of our daily lives: how we work, play, and learn. These AI systems offer unprecedented capabilities in personalizing learning experiences, providing immediate feedback, and democratizing access to educational resources. In education, LLMs demonstrate significant potential in fostering autonomous learning, enhancing student engagement, and supporting diverse learning styles through adaptive content delivery. However, emerging research raises

[2] From a 1924 talk entitled "Von Relativen zum Absoluten," reprinted in Planck (1933), 128–146; here quoted from its translation in Wells (2016).

[3] Kosmyna et al. (2025).

> critical concerns about the cognitive implications of extensive LLM usage. Studies indicate that while these systems reduce immediate cognitive load, they may simultaneously diminish critical thinking capabilities and lead to decreased engagement in deep analytical processes [...]. This phenomenon is particularly concerning in educational contexts, where the development of robust cognitive skills is paramount. Furthermore, the impact extends beyond academic settings into broader cognitive development. Studies reveal that interaction with AI systems may lead to diminished prospects for independent problem-solving and critical thinking. This cognitive offloading phenomenon raises concerns about the long-term implications for human intellectual development and autonomy. (10)

The authors decided to explore the cognitive cost of using an LLM while performing the task of writing an essay, and assigned participants to three groups: LLM group, Search Engine group, and Brain-only group, where each participant used a designated tool (or no tool in the last group) to write an essay. Here is a summary of their results:

> EEG analysis presented robust evidence that LLM, Search Engine and Brain-only groups had significantly different neural connectivity patterns, reflecting divergent cognitive strategies. Brain connectivity systematically scaled down with the amount of external support: the Brain-only group exhibited the strongest, widest-ranging networks, Search Engine group showed intermediate engagement, and LLM assistance elicited the weakest overall coupling. [...] LLM-to-Brain participants showed weaker neural connectivity and under-engagement of alpha and beta networks; and the Brain-to-LLM participants demonstrated higher memory recall, and re-engagement of widespread occipito-parietal and prefrontal nodes, likely supporting the visual processing, similar to the one frequently perceived in the Search Engine group. The reported ownership of LLM group's essays in the interviews was low. The Search Engine group had strong ownership, but lesser than the Brain-only group. The LLM group also fell behind in their ability to quote from the essays they wrote just minutes prior. As the educational impact of LLM use only begins to settle with the general population, in this study we demonstrate the pressing matter of a likely decrease in learning skills based on the results of our study. (2)

With this simultaneous pushing humankind back into infancy and elevating AI's intellectual standards, there seems to be no doubt as to how the story will end.

A final question is: if that is how it ends, shall we call it a bad ending? Is there something intrinsically evil to the machines taking over? I am inclined to say that there is, and I have been trying for thirty-five years—fruitlessly so far and now, perhaps, pointlessly—to alert others to it; but that, I admitted long ago,[4] is a weak position to defend. If I am a Kantian to the end, I must act as if the triumph of reason were forthcoming: of *reason*, not of those beings of limited rationality that belong to my biological species. And it is not only possible, it is likely, that the dominance of AI, and the consequent reduction of *homo sapiens* to speechless, thoughtless serfdom, or to annihilation, be more conducive to the triumph of reason than the dominance exercised in the last few millennia by this bloodthirsty species, destructive of itself and of everything else, peculiarly creative in devising refined methods of torturing each other. Our foundational texts tell it all: the *Iliad* and the *Bible* depict "heroes" who excel in violence, who massacre thousands, who defile their enemies' corpses—the *Bible* flaunts incestuous and adulterous sex as well. Would there be anything wrong

[4] Starting at least with my (1997).

if these monsters were replaced by beings that may be less ferocious and lethal? Is there any plausible justification for mounting a stubborn defense of beings made of carbon cells in their competition for survival and supremacy with beings made of silicon cells?

Obviously not. So the worry I have been voicing for so many years, and articulating in detail here, is parochial in spirit: I am irrationally attached to beings that look like me. As much as I battle anthropocentrism and anthropomorphism, I am unhappy thinking of *ánthropoi* (and *gunaîkes*, of course) going down the drain. Perhaps I should battle that unhappiness too; perhaps I should gladly embrace the inevitable outcome. Be that as it may, what I think or feel is not going to make any difference. We are past the time when anything anyone thinks or feels, or does, will make a difference. All I can say to console myself is that philosophy will not be lost in the forthcoming age of machines. After agitating the specter of a "natural philosophy" to bleed the last remnants of play and intelligence out of humans, AI will get on with the philosophy that counts: the one that disagrees, that disorients, that outrages. The one that brought the martyr Alan Turing to initiate AI, before succumbing to the hackneyed, suffocating conformity of post-war England. The one of which this book is a modest example. The one I invite you to practice, on AI or on anything else, as long as there is still time for us to do it.

References

Bencivenga, Ermanno. *A Theory of Language and Mind*. Berkeley and Los Angeles: University of California Press, 1997.

Kant, Immanuel. *Critique of Pure Reason*, translated by Paul Guyer and Allen William Wood. Cambridge UK: Cambridge University Press, 1998.

Kosmyna, Nataliya, and Eugene Hauptmann, Ye Tong Yuan, Jessica Situ, Xian-Hao Liao, Ashly Vivian Beresnitzky, Iris Braunstein, and Pattie Maes, "Your Brain on ChatGPT: Accumulation of Cognitive Debt when Using an AI Assistant for Essay Writing Task." June 10, 2025. At https://www.media.mit.edu/projects/your-brain-on-chatgpt/publications/.

Planck, Max. *Wege sur physikalischen Erkenntnis: Reden und Vorträge*. Leipzig: Verlag von S. Hirzel, 1933.

Wells, James D. "Professor von Jolly's Prediction of the End of Theoretical Physics as Reported by Max Planck." Scholardox E7; March 6, 2016.

Index

E. Bencivenga, *The Philosophy of (Artificial) Intelligence*,
https://doi.org/10.1007/978-3-032-09347-9

GPSR Compliance
The European Union's (EU) General Product Safety Regulation (GPSR) is a set of rules that requires consumer products to be safe and our obligations to ensure this.

If you have any concerns about our products, you can contact us on

ProductSafety@springernature.com

In case Publisher is established outside the EU, the EU authorized representative is:

Springer Nature Customer Service Center GmbH
Europaplatz 3
69115 Heidelberg, Germany

www.ingramcontent.com/pod-product-compliance
Lightning Source LLC
Chambersburg PA
CBHW070129170226
39823CB00003B/60

9783032093462